Flow

The Twists and Turns of a Life in Turbulence

Keith Moffatt FRS

Emeritus Professor of Mathematical Physics at
the University of Cambridge

πάντα ρει
Everything flows
Heraclitus, c. 600 BC

First published in Great Britain in 2025

Copyright © Keith Moffatt

The moral right of the author has been asserted.

All rights reserved.

No part of this publication may be reproduced, stored in a retrieval system, or transmitted, in any form or by any means, without the prior permission in writing of the publisher, nor be otherwise circulated in any form of binding or cover other than that in which it is published and without a similar condition including this condition being imposed on the subsequent purchaser.

Cover photo: © Jürg Alean.
A steam vortex ring emitted by eruption of Mount Etna, 24 February 2000

Design, typesetting and publishing by UK Book Publishing.

www.ukbookpublishing.com

ISBN: 978-1-917329-54-5

CONTENTS

FOREWORD vii

Chapter 1. Stepping Stones 1
Stresa 1960, Munich 1964, Stanford 1968, Moscow 1972, Delft 1976, Toronto 1980, Lyngby 1984, Grenoble 1988, Haifa 1992, Kyoto 1996, Chicago 2000, Warsaw 2004, Adelaide 2008, Beijing 2012, Montreal 2016, Milan 2020/21, Daegu 2024

Chapter 2. I am Preconditioned 23
Childhood, World War II, George Watson's Boys' College, Edinburgh University, Vacation jobs, RNVR, Trinity College Cambridge, Les Houches 1959, Assistant Lectureship, and Fellowship at Trinity 1961

Chapter 3. The 60s: Knotted Vortices 37
Tying the Proverbial Knot 1960, Marseille 1961, DAMTP, Zakopane 1963, The Exodus of Fred Hoyle, Move to the Press site, Stanford 1965, Johns Hopkins University, Strawberry Hill 1967, Helicity 1969, Student unrest 1969, Mean-field electrodynamics 1966/70

Chapter 4. The 70s: Dynamo Action 61
NATO Advanced Study Institute 1972, Potsdam 1972, Białowieża 1973, College Complications, Les Houches 1973, Trip to Lanestosa, ICTP Trieste 1974, Khartoum 1974, Beersheva 1974, Back to Trinity, Paris 1975/6, Bristol 1977-1980, Perugia 1978

Chapter 5. The 80s: Political Tangles 77
My fortuitous return to Cambridge 1980, Hester's lymphoma crisis 1982, USSR 1982, Head of Department 1983, Fergus: a doubly manic episode; Penelope in Nice and Susa 1983, Palermo January 1985, Lavrentiev Readings, Kiev 1985, Feynman and the first Dirac Lecture, China 1986, MADYLAM 1986/7,

UCSD 1987, Fergus deceased 1987, Patras 1988, Riga 1988, Austria 1989, The Abutilon miracle, IUTAM Symposium Cambridge 1989

Chapter 6. The 90s: The Ravelled Sleeve of Care 103
Tight knots, The Isaac Newton Institute, Batchelor's 70th birthday celebration, Tallahassee 1991, KITP Santa Barbara 1991, INI Dynamo theory 1992, Kyoto 1993, Croquet in Kyoto, Cargèse by car 1993, Ecole Polytechnique Palaiseau, SEDI at Whistler BC 1994, Gaulrig 1996, Director of INI 1996-2001, Skreen, Ireland 1998, Sir George Gabriel Stokes, Humphry Davy Lecture 1998

Chapter 7. The 00s: Toying with Spin 125
President of IUTAM,2000-2004, Euler's disc, The rising egg problem, The rattleback, An unsuccessful bid for funding, The Zakopane limericks 2001, Retirement 2002, Iran 2002, Blaise Pascal Chair 2002/2003, Yves Couder and bouncing droplets, The African Institute for Mathematical Sciences (AIMS), Muizenberg, South Africa, Botswana 2006, IMS-Singapore, Midwest tour 2003, Trinidad and Tobago 2004, ICTP Trieste 40th Anniversary, 70th Birthday, Université de Tous les Savoirs, Paris 2005, KITP Santa Barbara 2008, DAMTP Jubilee 2009

Chapter 8. The 10s: Topological Dynamics 153
Bozeman Montana 2010, Warsaw 2011, AIMS Senegal 2011, Topological dynamics 2012, NAS 2013, AIMS Ghana and AIMS Cameroon 2014, Konrad Bajer deceased 2014, Giffen Goods, 80th birthday, Research continued, Venice 2016, The finite-time singularity problem, Whittaker Colloquium 2017, APS Fluid Dynamics Prize and Otto Laporte Lecture 2018, Burns Night 25 January 2019, Oman 2019, Bailebeg Speyside

Chapter 9. The 20s: Pandemic Preoccupations 171
Timelessness and Eternity, The Pandemic, Barcelona 2024, Fingering instability, Frontiers in Dynamo Theory 2022, Miloška

Chapter 10. Paradise enow. 179
Wilderness were Paradise enow, Tam-day musings, The Genius o' Glenlair, On the Planting of Newton's Apple Tree, Dinner at Corpus, Genesis:Cosmological Echoes, A Natural Philosopher's Creed, GKB, Threescore Years and Ten, Part

II Limericks, Le Château de Tennessus, Black Swan: On First Looking into the Tswaing Meteorite Crater, To Sir Nicholas Barrington on his birthday, Bathsheba on her 14th Birthday, Address at our Golden Wedding Celebration, A Drinker's Guide to ICTAMs past

APPENDIX. Some Key Developments in Fluid Dynamics since 1956 211

JFM volumes 1-1000, Magnetohydrodynamics, Statistics of turbulence, Geophysical Fluid Dynamics, Dynamo Theory in Geophysics and Astrophysics, Microhydrodynamics and Suspension Mechanics, Biological Fluid Dynamics, Fractals and Chaos, Lagrangian Chaos, Direct Numerical Simulation (DNS), Coherent Structures, Concentrated vortices, Nonlinear stability, Convective turbulence, Free Surface Flows, REFERENCES

INDEX 225

FOREWORD

I realised only recently that I had attended more of the quadrennial Congresses of Theoretical and Applied Mechanics than anyone else in the world, fifteen in all since 1960. These Congresses have formed stepping stones across the river of my scientific life. I was encouraged to commit to paper some memories of these stepping stones before they are lost for ever. This has led to the content of Chapter 1 of this book. But there was much turbulence in the flow of the river between the stepping stones, so this writing of my memoirs has expanded to record some highlights of my life both before 1960 and in the subsequent decades.

The life of an academic is in a constant state of flux and tension between work and family life. The work itself is in perpetual tension between teaching, administration and research. Research can at times be all-consuming, and it is the occasional breakthrough in scientific research that lends excitement and a great sense of worth and achievement to this activity. I have recorded some of this previously on my website <www.keithmoffatt.com> (where more scientific detail can be found), but the time now seems ripe to create a more tangible, and perhaps more permanent record. This at any rate is my current aim.

In this endeavour, I have had the constant encouragement of my wife Linty; see the penultimate poem of Chapter 10 that reveals some highlights of our life together. Without Linty's unstinting support over the years, my efforts would have been fruitless. She has provided the life-work balance that will I hope shine from the pages of this book.

I have served the International Union of Theoretical and Applied Mechanics (IUTAM) in various capacities over a 50-year period, and in particular as President of IUTAM 2000-2004 and as Vice President 2004-2008. This provides the background to my scientific career. My research area has been fluid dynamics; I served as Editor of the *Journal of Fluid Mechanics* for 18 years, as Head of DAMTP Cambridge 1983-1991, and as Director of the Isaac Newton Institute, succeeding Sir Michael Atiyah, 1996-2001. These responsibilities have brought me into personal contact with legendary scientific figures including von Karman, Kolmogorov, Lighthill, Barenblatt, Bullard, Corrsin, de Gennes, Feynman, Hawking, Kovasznay, E.N.Parker, Kruskal, and many others including my own mentor George Batchelor. I devote a chapter to each of the decades from 1960 to the present day, and I include a preliminary chapter about my 'preconditioning' before 1960, and a final chapter with my poetry and some philosophical reflections. I include also an Appendix providing a very superficial account of some key developments in fluid dynamics since 1956, with which I have been particularly involved.

hkm
dec. 2024

CHAPTER 1

Stepping Stones

Introduction

I recently attended the International Congress of Theoretical and Applied Mechanics (ICTAM2024), held in Daegu, S. Korea. During the traditional banquet, I was awarded a prize for having attended more of the quadrennial ICTAMs than any of the other 2000+ participants present. This was in fact my 15th ICTAM; these have been the stepping stones of my scientific life, starting from Stresa, Italy (1960). The award of this prize, a beautiful piece of Korean mother-of-pearl craftsmanship, has stimulated me to write an account of my personal involvement with ICTAM, and its umbrella organisation IUTAM, over the past 64 years. Thank you 감사합니다 ICTAM2024!

The Congresses of Applied Mechanics were founded in 1922 on the initiative of Theodore von Karman, who wished to generate renewed international collaboration in the aftermath of the Great World War. The first Congress of Applied Mechanics was organised in Delft in 1924 by J.M.Burgers, with the involvement of G.I.Taylor as principal UK representative. The second was in Zurich 1926, and since then, they have been held every four years, with a hiatus during World War

II: Stockholm 1930, Cambridge UK 1934, Cambridge Mass.1938, then Paris 1946, quite remarkably, given the chaos that must still have existed in Paris after the end of the War. At the Paris Congress, organised by Paul Germain, the idea of establishing an International Union of Applied Mechanics was broached, and this was formalised at the subsequent Congress in London 1948. (This became the International Union of *Theoretical and* Applied Mechanics, IUTAM, somewhat later in 1972.) The Congress Committee of ICTAM has maintained strict autonomy, although under the umbrella of IUTAM.

Stresa 1960

In 1960, at 25 years old, I was a research student in Cambridge under the supervision of George Batchelor, eminent world authority in the subject of turbulence. He suggested that I should attend the 10th ICTAM, held in Stresa, Italy, on the beautiful shore of Lake Maggiore. I did so with alacrity in the company of Dick Jarvis, a fellow research student in Cambridge, memorable for the quantity of vermouth that he consumed at the Welcome Reception. Von Karman himself was present at this, his last, Congress (he died in 1962); he was also present at the *Colloque International sur la Mécanique de la Turbulence* held the following year in Marseille; this was a more compact gathering of about 60 participants with a single session of lectures every day for a week. I recall that von Karman would arrive in the lecture hall about five minutes late every morning, when the first

Batchelor (right) in discussion with von Karman, Marseille 1961

lecturer of the day was just launched into his presentation. He, von Karman, would come slowly down the central aisle shaking hands warmly with those on both sides while we all stood in recognition of his eminence and authority. This arrival ended with von Karman embracing G.I.Taylor and settling down with him in the front row; only then could the distracted lecturer resume his presentation. I shook hands with von Karman on this occasion, making me one of the few surviving links with the origins of ICTAM and hence of IUTAM.

Anyway, Stresa 1960 was my first experience of ICTAM. In those days there were four 'General Lectures', by invitation from eminent authorities. One of these was by Bogoliubov on *"Méthode analytique de la théorie des oscillations non-linéaires"*; another by J.T. (Trevor) Stuart on *"Nonlinear effects in hydrodynamic stability"*, a subject in which he was a pioneer. All other lectures (of which there were 184 in several parallel sessions) were 'contributed lectures', which had been subjected to a rigorous selection procedure. One of these was by G.I.Taylor on *"Waves on thin sheets of fluid"*, a characteristically innovative topic at that time. There were many further eminent names in the list of 742 participants; among these (in no particular order): George Carrier, Imai, Hans Liepman, Naghdi, Sedov, Zorski, Barenblatt, Favre, J.B.Keller, Lagerstrom, Loitsianski, Schlichting, Mushkelishvili, Milton Van Dyke, O.M.Phillips, Budiansky, den Hartog, and Gerry Whitham; none alas alive today, but what a feast it provided then for research students privileged to participate in such a Congress!

Munich 1964

At the Closing Ceremony, as has been normal from the beginning, the location of the next congress was announced: Munich 1964. Also, among other announcements, Batchelor was appointed Treasurer of

IUTAM, an association that inevitably drew me also into this sphere of activity. I attended Munich 1964 (and contributed a lecture there on the subject *"Electrically driven steady flows in magnetohydrodynamics"*). I travelled to Munich with Anthony [J.R.A.] Pearson in his camper van; there were two others in the party, Trevor Waechter and a Scot from Dundee whose name I forget, but, with his strong Scottish accent, he was naturally known to us simply as Jock. One incident from that journey remains vivid in my mind. We camped somewhere in the French Jura, and Jock set to work to inflate Anthony's new tent, of which he was visibly proud. This had no metal framework, but relied on inflatable tubes built into the ribs of the tent --- quite novel at the time. Jock sat inside the growing structure and pumped with great vigour; too great, because with a loud bang the tubes exploded and the tent gently subsided on Jock who remained inside, seemingly himself quite deflated. Anthony was understandably crestfallen at this turn of events! Nevertheless, he delivered us eventually to the Congress location in Munich.

But there our adventures did not end, for we naturally spent our evenings in the thriving bars of the city. In one of these, certain very glamorous maidens draped themselves around our party of four, and ordered several bottles of champagne, which they rapidly consumed. When we rose to leave, the bill included the greatly inflated cost of the champagne, which Anthony, who spoke good German, in a rage refused to pay. A row ensued and within minutes armed police arrived on the scene and carried Anthony off to the cells where he spent a comfortable night and was then released without charge. Anthony, now 94 years old, has vouched for the veracity of this account.

I recall also that somehow Anthony and I engineered an invitation to (i.e. gate-crashed) a dinner hosted by Clifford Truesdell, already famed as the founder of the *Archive for Rational Mechanics and*

Analysis. At this dinner, I met Grisha Barenblatt, who was to play an important part in my scientific life much later.

Stanford 1968

Then there was the Congress of Stanford 1968, hosted by Nicholas Hoff. It is worth noting that until this time there were four official languages, English, French, Italian and German, in any of which participants could present their lectures; nearly all chose to lecture in a sort of 'international' English, since otherwise their audience was sparse. There was some consternation in Stanford however, because it emerged that the French delegation were under strict orders to present their lectures in French, in a last-ditch attempt to maintain the French language, in parallel with English, as an internationally accepted language of science. The French participants were themselves unhappy about this 'directive from on high', but were obliged to conform. By the time of the Delft Congress in 1976, English was firmly established as the natural

Sweet Emma and the Old Preservation Hall Band, 1968

language of ICTAM, although, as far as I know, the legitimacy of French, Italian and German has never been formally repealed. [Regarding this 'international English', I recall overhearing a Fellow of Trinity asking Abram Besicovich, also a Fellow, why, having lived for many years in Cambridge, he still spoke with a strong Russian accent. He replied *"English you speak, only English speak; English I speak, half world speak."*]

Milton Van Dyke was responsible for the social programme at the Stanford Congress, and had succeeded in attracting the Old Preservation Hall Band from New Orleans (all then in their 70s) to come to Stanford, where they provided a sensational evening concert of traditional jazz. Milton held a party for the musicians at his home the following evening, when I was able to meet these wonderful old-timers. I bought their record which I treasure to this day.

Moscow 1972

This was the only ICTAM since 1960 that I was unable to attend, my father being very ill at the time, and our four children, now aged 10, 9, 6 and 4, being all at an exciting and demanding stage of development. I therefore skipped this stepping stone; my student Rick Dillon was however able to attend, and reported back that it had been uncomfortably hot, well over 30°C, throughout the week of the Congress. I was comforted by the fact that I had visited Moscow in 1965 for the International Colloquium on *Atmospheric Turbulence and Radio Wave Propagation,* and still cherished happy memories of that earlier visit.

Delft 1976

Some years later, I was appointed to the International Papers Committee (IPC) for the Congress to be held in Delft 1976. The Committee met for four days in April 1976 under the strict chairmanship of Warner Koiter. I worked with Leen van Wijngaarden on the fluids side. We read all the submitted abstracts, taking due account of National Committee recommendations, and agreed the final selection within the four-day limit; it was very hard work! Koiter was a tower of strength in every respect.

On Koiter's recommendation, I was appointed to the Congress Committee (CC) and to its executive sub-committee (the XCC) some months in advance of the Delft Congress. I served on the XCC till Haifa 1992, and then continued as a member of the Congress Committee and of the General Assembly.

In 1978, a combined meeting of the General Assembly of IUTAM and the Congress Committee was held in Herrenalb (BRD). Being invited to give a lecture at this meeting, I presented my work on helicity invariance and associated dynamo action. Batchelor told me that this excited much interest among the Russian delegates, led by L.I.Sedov. [Lighthill also expressed interest in my presentation, when I was then able to expand on the seminal work of the Potsdam group, Steenbeck, Krause and Rädler.]

Toronto 1980

In April 1980, again with Leen van Wijngaarten, I was still on the IPC selecting papers for Toronto 1980. This period was fraught for me personally, because my elder son Fergus, aged 19 and later diagnosed as bipolar, was hospitalised following his first manic episode. I recall telling James and Nancy Lighthill about this, while waiting for the plane home following this April meeting. They recognised how serious this situation could be, having possibly encountered similar problems themselves. More on this later. I have little recollection of the Congress itself in August that year.

Lyngby 1984

In 1982, the General Assembly had its 'intercalating' meeting in Trinity College, Cambridge, UK, at my invitation. The Congress Committee was now planning for Lyngby 1984. Bruno Boley was Secretary of the XCC, but was unable to attend due to serious illness of his wife; I stood in for him on this occasion. [This meeting was followed immediately by the IUTAM Symposium "*Metallurgical Applications of Magnetohydrodynamics*" which had been proposed by Arthur Shercliff, Head of Engineering in Cambridge; sadly, Arthur had been stricken earlier that year with cancer, and was in hospital at the time of the Symposium, for which Michael Proctor and I assumed responsibility, reporting daily to Arthur who remained keenly interested. Arthur tragically died from this cancer one year later, aged 56.]

James and Nancy Lighthill with Grisha Barenblatt (left), Cambridge 1991

The Lyngby Congress, under the commanding Presidency of Frithiof Niordson, was innovative in two respects: the inclusion of poster/discussion sessions, and the inclusion of three mini-symposia on certain well-defined sub-topics. These developments were generally welcomed and have been an important ingredient of all subsequent ICTAMs. I recall Frithiof's entertaining speech at the Banquet, which included his immortal adaptation of Shakespeare,

with Frithiof Niordson (on left) and Niels and Mrs. Olhoff, at the Cambridge GA, 2002

"*There's something rotten in the State of Denmark, but I hope it's not on the menu tonight*"!

Grenoble 1988

A meeting of the General Assembly of IUTAM and of its Congress Committee was held in London in 1986, at the invitation of James Lighthill, then President of IUTAM; I was present at this meeting, as Secretary of the XCC. The meeting was of historic importance because two delegates from China were present for the first time: Zheng Zhemin and Wang Ren. China had been admitted to membership of IUTAM in 1980 following the death of Mao Tse Tung in 1976; it would play an increasingly active role in the affairs of IUTAM from this point on.

In September 1987, a meeting was scheduled in Paris to settle the final announcement for Grenoble 1988. As Secretary of the XCC, I had worked on this for some months, and had a final draft ready, with all supporting papers. Just two days before this meeting, my son Fergus, who had been in a deep depression, took his own life. This was a crisis situation for my whole family, but Paul Germain (*Secrétaire en perpetuité* of the Académie des Sciences), James Lighthill and Bruno Boley were already awaiting my presence in Paris, and I just had to be there two days later. I made the trip to the Académie with a heavy heart, but the essential planning for Grenoble 1988 was then successfully accomplished and finalised.

While in Paris, a miracle occurred. My elder daughter Hester, then a student of French and Modern Greek at Oxford, had set off a week earlier for the University of Thessalonika, where she was to spend a year in part fulfilment of her course requirements. We had tried repeatedly, but unsuccessfully, to contact her by phone the previous day to tell her the tragedy of Fergus's death, and to ensure that she could return for the funeral. That evening, in my hotel in Paris, I opened a Gideon bible at random: it fell open at St Paul's First Epistle to the Thessalonians, where I could swear I read the lines "*I called unto Thessalonika and ye answered me*". I called immediately unto Hester, and she answered me!

During Grenoble 1988, the Congress Committee had to decide on the location for the 1992 Congress. The meeting at which this decision was taken was quite dramatic. There were two leading contenders, China (Beijing) and Israel (Haifa). Lighthill, in his Presidential capacity, favoured Beijing, but he overplayed his hand and his bias irritated some members of the CC, with the consequence that the final vote favoured Haifa by a narrow majority. [Political unrest in China a year later led many to consider that this was a fortunate outcome. Beijing was to succeed in more settled times -- Beijing 2012].

David Crighton and Hassan Aref at the IPC meeting in Haifa, March 1992

Haifa 1992

The decision to hold the Congress in Haifa was equally problematic. The Congress Committee met in Vienna on 31 August 1990, and as Secretary of the Committee, I wrote following that meeting: "*At the time of writing this report, the Gulf War is at a critical stage, and there must be some uncertainty as to whether the political situation will in fact permit the holding of the Congress in Israel in August 1992. The Executive Committee is considering the options with great care and will take any decision that the prevailing situation may demand, when it meets in Udine, Italy, 31 August and 1 September 1991*". In the event, the political crisis had by then abated, and the arrangements for Haifa 1992 went ahead as planned.

The International Papers Committee met in Haifa in March 1992. I was present as Secretary and Hassan Aref and David Crighton were the 'fluids' members; they generously allowed my (late) submission of a paper (with Jae Tak Jeong) on: *"Free surface cusps in flow at low Reynolds number"*. The Haifa Congress was small-scale (525 participants from 40 countries), but high in quality. I was elected to the Bureau of IUTAM by the General Assembly meeting during this Congress.

Kyoto 1996

IUTAM Bureau 1996-2000, from left: M.A.Hayes (Ireland, Secretary-General), T.Tatsumi (Japan), HKM (UK), J.Engelbrecht (Estonia), W. Schiehlen (Germany, President), L. van Wijngaarden (Netherlands), L.B.Freund (USA, Treasurer), Wang Ren (China)

The decision in 1992 to hold the following Congress in Kyoto (Japan) 1996, the first to be held in Asia, was met with general excitement and enthusiasm. And it was very fortunate for me personally, because I was already engaged to spend the period, January to March 1993, at

RIMS (the Research Institute for Mathematical Sciences) in Kyoto, at the invitation of Shigeo Kida. My wife Linty and I spent a delightful three months in Japan, visiting not only Kyoto, but also Sapporo (Hokkaido) in the north, and Fukuoka (Kyushu) in the south, staying in Japan just long enough to see the famed flowering of the cherry blossom in Kyoto in early April.

With Tatsumi, Kambe and Sano before the Great Buddha in Tara, south of Kyoto

I returned to Kyoto at the end of May for the meeting of the XCC, to view the Congress facilities and to plan for Kyoto 1996, for which Tomomasa Tatsumi was President of the Local Organising Committee (LOC). Tatsumi had been a Senior Visitor to Cambridge in the mid-1950s, the first Japanese visitor to Batchelor's research group after the War. He told me that his visit coincided with that of Judge Hisashi Owada, father of Masako, current Empress Consort of Japan. Owada is an Honorary Fellow of Trinity College, Cambridge. Tatsumi entertained us royally at this meeting in Kyoto, including a visit, very properly conducted, to a Geisha parlour in downtown Hanamikoji; very quaint! Tatsumi died in 2023 at the age of 100, while still pursuing his life's work on his closure theory of turbulence!

The Congress duly took place as planned in Kyoto 1996, when I was serving on both the Bureau of IUTAM and the XCC. The photo shows the Bureau elected in 1996. [It is sad to note that Michel Hayes of the University of Limerick, who then served as Secretary-General, died tragically in a fire at his home on the Limerick campus in June 2024; he was 91.] Three of my students, Paul Dellar, Vincent

Mak and Atta Chui, had lectures accepted at the Kyoto Congress. It is interesting to note that the use of transparencies in lectures had by then become the norm rather than photographic slides; [this did not last long, with the increasing accessibility of PowerPoint during the 1990s!]. Lighthill gave the Closing Lecture *"Typhoons, Hurricanes and Fluid Mechanics"*; his transparencies were densely packed with hand-written scientific detail, and his style of presentation was magisterial, with little respect for the constraint of time, somewhat to the dismay of our Japanese hosts, who had arranged the spectacular Closing Ceremony that followed with meticulous attention to every detail. Lighthill died just two years later at the age of 74 while swimming round the Channel Island of Sark, a feat that he had accomplished previously, but which this time proved his nemesis; as I wrote in my *Nature* obituary of Lighthill, *"he died as he had lived, with style and bravado"*.

Refreshments in my room at the Kyoto Congress 1996; from left: standing: HKM, Yoshi Kimura, Renzo Ricca; sitting: Gary Leal, Volodya Vladimirov, Osamu Sano; reclining: Paul Dellar

Chicago 2000

Hassan Aref was President of the LOC for the 'Millennium Congress', Chicago 2000. The Bureau and XCC held its 1997 planning meeting in Urbana, Illinois, in the course of which Hassan invited us all out to his home in deep farming countryside, some 25 miles west of Urbana --- a memorable visit.

At the Opening Ceremony, in recognition of the historic character of the occasion, Aref had engaged a number of actors to portray an imaginary discussion between Archimedes, Galileo and Newton, a poetic and colourful start to the Congress, which was nevertheless found by some to be a bit perplexing! A group of us entertained Stephen Wolfram (creator of *Mathematica*) to dinner in the Congress Hotel one evening; he talked nonstop, and was very opinionated.

Aref declared Post-Congress that the 27th day in August should henceforth be declared "TAM-day" (for **T**heoretical and **A**pplied **M**echanics); this led me to compose my satirical poem "*TAM-day musings*" (see Chapter 10). The Congress progressed well, and I gave the Closing Lecture on "*Local and global perspectives in fluid dynamics*". At the meeting of the General Assembly, I was elected President of IUTAM for the next 4 years, an elevated responsibility that I was both honoured and privileged to accept.

Warsaw 2004

I was particularly glad too that Warsaw was chosen for the 2004 Congress. Poland had served as a meeting ground for fluid dynamicists from the USSR and the West during the difficult cold-war years from 1960 to 1980, when such contact was difficult to maintain, and further during the development of

From the left, Batchelor, Fiszdon and Herczynski in Blazejewko 1979

Solidarność during the 1980's. Throughout this whole period, George Batchelor had been active in supporting the biennial meetings in fluid mechanics for which Wladek Fiszdon and Ryszard Hercynski were the prime movers. I had attended two of these meetings myself in Zakopane (1963) and Białowieża (1973), and had good reason to be grateful for the interactions that such meetings provided.

So I approached the 2004 Congress with keen anticipation. The Bureau and XCCC had its first planning meeting in Warsaw in August 2001, and I returned to Poland (with Linty this time) for an IUTAM Symposium, "*Tubes, Sheets, and Singularities in Fluid Mechanics*" held in Zakopane just one month later. This was organised by Konrad Bajer, my former student and close friend. In August 2002, the General Assembly met at the Møller Institute in Cambridge at my invitation.

In September that year, I represented IUTAM in Rio de Janeiro at the General Assembly of ICSU (the International Council for Science); Juri Engelbrecht was also there, representing Estonia, and we spent social time together. By lucky chance, the IUTAM Bureau was scheduled to meet in Tallin the following year. There, at Juri's instigation, we were entertained at the British Embassy by Sarah Squire, at that time British Ambassador in Estonia.

At the General Assembly meeting during the Warsaw Congress, I was able to report the award of an ICSU grant of $100,000 to IUTAM in support of the capacity-building initiative "*AIMS: African Institute for Mathematical Sciences*". This grant provided support for a Workshop on *Capacity-Building in the Mathematical Sciences* held at AIMS, Muizenberg, South Africa, in April 2004, which I had attended, together with Hassan Aref, Dick van Campen and Jean Salençon. AIMS had been founded in 2002 by Neil Turok, at that time Professor of Cosmology at the University of Cambridge. Neil had consulted me from the beginning, in seeking to replicate the success of the Isaac Newton Institute, and I was glad to be able to serve in an advisory capacity on the Council of AIMS for the first 15 years of its existence.

In October 2005, I again represented IUTAM at the General Assembly of ICSU, this time held in Suzhou, China, an ancient city of "canals, stone bridges, pagodas, and meticulously designed gardens". It should be twinned with Cambridge! The participants were given a parting gift of a beautiful and well-filled, special limited-edition stamp album, a most unconventional souvenir.

In July the following year, Linty and I spent several weeks at the Woods Hole Oceanographic Institution, on Cape Cod, before going on to a meeting of the General Assembly of IUTAM in Providence, RI, hosted by Ben Freund (1942--2024), who had succeeded me as President of IUTAM. Alex Zaslavsky paid this tribute to Ben, who died on 3 October 2024: *"Not only a great scientist, but an unusually kind and generous person, as everyone who has ever interacted with him will remember"*. And I do indeed remember Ben with great affection.

Adelaide 2008

IUTAM Bureau 2004-2008, from left:
A.Kluwick (Austria), T.Kambe (Japan), HKM (UK), D.H.van Campen (Netherlands), Z.Zheng (China), L.B.Freund (USA), J.Engelbrecht (Estonia), N.Olhoff (Denmark)

At the Closing Ceremony of the Warsaw Congress, I announced that the 2008 ICTAM would be held in Adelaide; this was the first ICTAM to be held in Australasia. [Adelaide had historic interest for ICTAM in that Horace Lamb had spent some years at Adelaide University when he was writing the first edition of his *Hydrodynamics*, the bible of the subject until at least 1960.] Ernie Tuck was appointed as President of the Local Organising Committee. Ernie had taken his PhD in Cambridge under Fritz Ursell, who succeeded Lighthill in Manchester in the early 60s. I knew Ernie from these early days. In 2005, when the Bureau and XCC held their planning meeting in Adelaide, Ernie and his wife took me on a tour of the vineyards in the hills of the hinterland; we sampled some fine wines of the region! Sadly Ernie was stricken with cancer early in 2007, but bravely fulfilled all his responsibilities as Congress President in 2008; he died in March 2009, aged 69.

The Proceedings of the Congress were somewhat delayed but eventually appeared in 2013, under the title *"Mechanics Down Under"*. I have a paper in that volume with title *"Relaxation to steady vortical flows, and knots in the quark-gluon plasma"*, this being as near as I have ever got to elementary particle physics. [The delay in publication brought to mind the Paris Congress of 1946, for which the Proceedings, according to Paul Germain, were duly delivered to the Publisher, Gauthier-Villars, but have not yet appeared in print!]

With Slava Meleshko (centre) and Hassan Aref at a meeting on vortex dynamics, Moscow 2006

Following the Adelaide Congress, once again I represented IUTAM at the ICSU General Assembly held in Mozambique in October that year. I made a presentation on behalf of IUTAM and focused largely on the AIMS project that was now supported by IUTAM as well as five other International Scientific Unions. This led to a meeting with the Minister for Science and Technology, Venâncio Massingue (1960-2017), an impressively authoritarian figure. He slapped the table and said "I want an AIMS in Mozambique" and instructed his charming assistant Sarifa Sagilde to liaise with me in the matter. I phoned Neil Turok to tell him the good news, but alas, nothing came of this in the end, when Massingue learnt that Mozambique would have to provide the building to house an AIMS Institute, and also contribute to the funding. My evenings in Maputo were enlivened by my interactions with John Buckeridge, President of the International Union of Biological Science (IUBS), over drams of the Talisker that I had had the foresight to buy in Duty Free!

Beijing 2012

Beijing was ultimately successful in its bid for the 2012 Congress, much to the delight of the Chinese members of the General Assembly. The Congress took place very close to the site of the 2008 Summer Olympic Games, and miraculously the weather was beautiful and the air clear and completely free throughout the week of the dust and pollution that had been widely anticipated. Hassan Aref had been appointed Secretary of the Congress Committee at the Adelaide Congress, and served in this capacity until his sudden tragic death on 9th September 2011 at the age of 60 from aortic dissection. The photo shows me with Hassan and his colleague Slava Meleshko from Ukraine, taken during a meeting in Moscow on vortex dynamics. In further tragedy, Slava (known as a 'walking encyclopedia' in the history of vortex dynamics) was killed just two months later, hit by a bus in Kiev, while crossing a road.

Aref's responsibilities for the Beijing Congress were subsequently shared till 2012 by Tim Pedley, Dick van Campen and Viggo Tvergaard.

Montreal 2016

One highlight at the Montreal Congress in August 2016 was the Batchelor lecture given by Ray Goldstein on the subject *Fluid dynamics at the scale of the cell*. I had been collaborating since 2010 with Ray and his wife Adrianna Pesci (and also with Renzo Ricca) on the fascinating

With Krzysztof Mizerski and Yoshi Kimura at ICTAM Montreal 2016

problem of the collapse of a 'one-sided' soap film in the form of a Möbius strip, so was delighted that Ray was the 2016 winner of the Batchelor Prize.

Following the Congress, I was able to visit my Auntie Phoebe in Ottawa, who had reached the venerable age of 104. She was in a care home, but was still amazingly alert and articulate as we poured over old photos from her early (pre-war) life in Scotland. I met many cousins and their families in Ottawa during that visit. Phoebe died just one month later, by which time I was back in Cambridge.

Milan 2020/21

The Milan Congress was, like so much else, devastated by the Covid pandemic; it was first postponed to 2021, and then the difficult decision was taken to run it entirely online by Zoom. I had registered and was very much looking forward to this Congress, but I find it very difficult to concentrate on Zoom lectures, so I finally withdrew. I resigned from the General Assembly in 2023, having served on this and/ or the Congress Committee for 47 years; it was clearly time to make way for the younger computer-literate generation!

This leads me back full-circle to

Daegu 2024

Here, my younger daughter Penelope was my 'accompanying person' and I was indeed very glad to have her company. For the first time ever, I prepared a poster, having enjoyed poster sessions at many other meetings. We were given 3 minutes to summarise, and in effect advertise, the content of our posters; I was lucky in that the two following speakers did not show up, so I had six more minutes, giving ample opportunity for questions and discussion. I really enjoyed the occasion, and was able to include my summarising limerick: *Two vortices stretch and deform/To a length far exceeding the norm;/Our dynamical system/Shows well how to twist 'em/And certainly won't misinform!*

Dinner in the karaoke restaurant, Seoul, where aprons were provided for cooking at the table

We had flown by Korean Air direct to Seoul, where we stayed two nights. We visited the Changdeokgung Palace of the Kings of the Josean dynasty (1392--1910), with its 'Secret Garden', no longer quite so secret. And we dined that evening in what turned out to be a karaoke restaurant, where Penelope sang "*A Hard Day's Night*" to rapturous applause.

In Daegu, the Wednesday excursion took us to the Donghwasa Temple, of the the Jogye Order of Korean Buddhism (대한불교조계종, 大韓佛敎 曹溪宗) high up on Mt. Palgongsan.

Pause for breath with Martin Oberlack at the Donghwasa Temple

The banquet was expeditiously delivered to the myriad of round tables, course by course, so expeditiously that my main course was whipped away while I turned my back to our table in order to watch the brilliant Taekwondo demonstration that was taking place on the stage. It was during this banquet that the President, Norman Fleck (of the Cambridge Department of Engineering), determined by a process of successive elimination that I was indeed the participant who had attended more ICTAMs than anyone else, thus qualifying me for the 'longevity' prize (as I like to describe it).

The award of this prize stimulated me to record in a *'Drinker's Guide to ICTAMs Past'* (see Chapter 10) some of the beverages that helped to sustain participants through each of these historic ICTAMs.

CHAPTER 2

I am Preconditioned

Childhood

I am a Scot on both sides of my family, born and bred in Edinburgh, under the gorse-clad slopes of Blackford Hill. My earliest memories are of the War years and of the fireworks when it all came to an end. I recall gas masks, bombers overhead on their way to and from the Glasgow docks, and the absence of bananas, for which I had a peculiar craving, having tasted one before the War.

I joined the cubs at the age of 8 and learnt how to sew on buttons, tie reef knots, bowlines, clove hitches, and sheet bends, skip a hundred times, and cook bannocks over an open fire, skills that have stood me in good stead throughout life.

My mother was descended from a Highland clan that lived in the clachan of Gaulrig, about a mile south of Tomintoul. Little remains of Gaulrig except for a few tumbledown cottages that the inhabitants used to share with their livestock. In 1820, my great great great grandmother, who had been widowed for five years, died in dramatic circumstances. Family chronicles relate that she was burnt to death in an illicit still. I imagine that she was forewarned of the approach of the dreaded Excisemen, and so set fire to her whisky still in order to destroy the evidence of this illegal activity. This was a hazardous procedure because of the inflammability of the product, and she would have perished in the resulting conflagration. I am comforted by the thought that the fumes of whisky filling the air may have provided solace in her dreadful plight.

My parents, Fred and Emmeline, at Strawberry Hill, Jamaica, December 1967

My father's family were on the consuming, rather than the production, side of the whisky industry. This meant that my parents' partnership was based on sound economic principles, if somewhat fraught at times. However, it worked out well in the end: quite late in life, following a retirement cruise in the Caribbean, they bought Strawberry Hill Hotel in the Blue Mountains of Jamaica. They spent six happy years there, just when Bob Marley and his reggae music were gaining popularity. They sold the hotel in 1972 to Chris Blackwell, founder of the Island Records label, and it became a centre for music of an avant-garde variety.

My father taught me an early love of numbers; he also taught a range of other life-enhancing skills, such as the ability to recite the names of the 33 counties of Scotland, from Shetland in the north to Berwickshire in the south. Don't get me started!

World War II

I was just four years old when World War II was declared. I remember that very day, the 3rd of September 1939: we were on holiday at Dirleton, a seaside resort in Berwickshire, when my sister Lindesay and I were mysteriously spirited away to live with a family in the village of West Linton, some 15 miles outside Edinburgh. In other words we were evacuated from the city, as many children were, such was the fear of enemy bombardment. I apparently reacted badly to this experience, so was soon returned to my mother in Edinburgh. My father had by then been called up, and I saw nothing of him for the next 6 years. Lindesay was evacuated for a second time to live with her paternal grandmother in Lasswade; our encounters during the war were infrequent, and all the more memorable on the rare occasions when visits were possible.

Granny Fleming
[1878--1976]

Throughout most of the War years, my mother was in the WAAFs, and much distracted by the war effort, and my maternal grandmother, Granny Fleming, cared for me. This grandmother was a gifted pianist, a lover of Chopin and Tchaikovsky; musical talent has passed me by, but Lindesay has this talent in abundance, as do all the younger members of our family.

In the event, Edinburgh was spared bombardment, though we frequently heard the bombers flying overhead to deposit their cargo on the Glasgow dockyards. I remember being put to bed in the bottom drawer of a large kitchen cabinet on one such occasion; it was all very exciting. We of course had gas masks, ration books, one egg each per week, no bananas, and strictly enforced blackout at night. I remember an emotional presentation at my first primary school about the siege and courageous defence of Malta in 1942. At this school, I learnt the history of the Scottish kings from Robert the Bruce to James the 6th and 1st, which was where history stopped. I also became a wizard at mental arithmetic, which served me well in later life. I was also an inveterate liar; I remember being led by the teacher to the washroom to have my tongue washed in order to cure me of this habit, but this alas had little salutary effect.

George Watson's Boys' College

At age eight, I moved to George Watson's Boys' College, a Merchant Company School that taught us where to place apostrophes, where I remained for the next ten years. I found myself in a class of boys that was exceptional, although I didn't appreciate this at the time: one (Jim Hiddleston) became an expert on Baudelaire and tutor in French at Exeter College Oxford, another (John Sawyer) became Professor of Divinity at Newcastle, a third (Eric Anderson) became Headmaster of Eton, and a fourth (Fergus Craik) achieved distinction for his research on the process of memory -- he was elected to Fellowship of the Royal Society for this work some years ago; and so on! Such company made for a competitive environment, from which I suffered no harm. I did however suffer harm on the rugby field, where, being small for my age and not particularly agile, I played hooker in the 3rd fifteen, and regularly found myself crushed under a collapsing scrum! The game of golf was more to my liking.

Scottish education was broad, and, although mainly in the science stream, I was able to continue with French, English, Latin and History until taking the Scottish Higher Leaving Certificate at age 17. We then had a further year at school preparing for the Edinburgh Bursary Competition. My final term at Watson's was enlivened by an exchange with the Lycée Henri IV in Paris. There were five of us, clad in kilts, who went to Paris for the term, exchanging with five French boys who went the other way, each living *en famille*. My host family, la famille Gama, lived in an apartment on the 6th floor at 124 Boulevard Raspail, where there was a hydraulic lift --- you pulled a cord in the lift and up it went, clunking all the way. From there, I walked each morning across the Jardin de Luxembourg, up the Rue Soufflot, and to the Lycée Henri IV behind the Panthéon. Remarkably, wine was served at the school lunches there, making this a spirited learning experience. I don't remember the afternoon classes because we played truant much of the time in the cafés of the BouleMiche.

For a boy who had been raised under strict Presbyterian control, this was a liberating experience. The redeeming feature of my Protestant upbringing was that the Minister of the Parish Church of Inveresk where we lived had a daughter named Linty, the youngest of a family of seven, who dazzled me from the age of 13. What dazzled me most was that she could dive from the 10m diving board at the huge open-air Portobello swimming pool, a feat that I could never equal. I resolved there and then that I would marry this talented girl at the earliest opportunity.

Edinburgh University

At Edinburgh University, I was lucky to be taught by the legendary Professor A.C.Aitken, famed for his feats of numerical dexterity in the multiplication of large numbers — he could still beat the primitive electronic computers of his day in this. He told us of his experience at Gallipoli in the Great World War. when all his platoon records had been destroyed and many of the platoon killed in enemy conflict. No problem, he had memorised the names and numbers of all 56 soldiers in the platoon, and the records could be faithfully restored.

I recall running into Professor Aitken some years after leaving University, at the harbour of St Abbs, Berwickshire, where he was watching a group of young fishermen competing with each other in attempting to throw stones right across the harbour to hit the wall on the other side. Aitken, then in his 60s and looking very much the elderly professor, picked up a smooth stone and prepared to join the contest; the fishermen looked on in some amusement. Aitken swung his arm vigorously in a highly unconventional way, and released the stone in an underarm throw; it hit the far wall to the amazement of the onlookers. He explained that he had adopted the principle of conservation of angular momentum, treating his arm as a compound pendulum!

On the Applied side, I was taught by Robin Schlapp and Andrew Nisbet; also by Nicholas Kemmer (who succeeded Max Born in 1953 as Tait Professor of Mathematical Physics). Kemmer lectured without notes, giving the impression that he was creating the subject from first principles with every lecture. I resolved to follow this inspiring example throughout my own lecturing career some years later.

There was also W.L.Edge, who taught geometry over finite Galois fields. He was a keen hill walker, and had accumulated a following of alumni and current students who, on an April day every year, climbed Stùc a' Chròin, a Munro in Perthshire. I was privileged to join this 'Stùc a' Chròin club'. Everything was 'canonical' about these trips: an early train from the old Caledonian station to Calendar, breakfast in the Crown Hotel, a long trek to the summit of Stùc a' Chròin, then through the saddle point to the summit of Ben Vorlich, descent to Edinample and then the hike back to Balquhidder to catch the last train back to Edinburgh -- a long testing day!

The Stùc a' Chròin club': Edge on extreme right, I am behind him, Robin Schlapp centre right, with Charlie Glennie, behind him; Ian Porteous crouching at the front.

Edge was intensely loyal to Trinity College, Cambridge, where he had been a research Fellow in the 1930s, and he induced a number of his students, including Ian Porteous and Charles Glennie, to take the Trinity Entrance Scholarship examination. He didn't hold out much hope for me, because geometry was not my strong subject, but he encouraged me to try for the Trinity group of Colleges, with the grudging words *"You may get into Magdalene, and Babbage will be happy to teach you"*. I remember staying in F1 New Court when I came to take the Scholarship exam; it was sheer bliss! In the event, I got a minor scholarship to Trinity, worth £60 a year, a fortune in those days. My Edinburgh classmate Jim Mirrlees won a major scholarship (£100) on the same occasion, and we both matriculated in October '57 as affiliated students to read Part II of the Mathematical Tripos.

Jim had managed to skip the first year at Edinburgh, and came straight into the second-year class. In the 3-hour exam at the end of his first term, he walked out after 2 hours, and we all thought *Poor Jim, he can't solve any of these difficult problems*; it turned out that he had solved them all, and saw no point in staying longer than necessary! From that point on, his Nobel Prize (in Economics, 1996) was never in doubt.

Vacation jobs

While still at Edinburgh, it was necessary for me to seek gainful employment during University vacations, in order to make ends meet. One of my vacation jobs was on night shift, 10 p.m. to 6 a.m., at the Walls' Ice Cream factory outside Edinburgh. My task was to stand at a machine that chugged out those awful rectangular-wrapped ice-creams on a conveyor belt, pick up six of these in my left hand, transfer them to my right hand, and place them on a second conveyor belt at right angles to the first, where they were machine-wrapped in packets of six, and carried away, I knew not where. I became literally a dab hand at this process, but after several hours, the novelty wore off. The only way to get an unscheduled break was to drop an ice-cream into the chugging machinery, Luddite manner, thus bringing it to a grinding halt!

At the age of 18, I passed my driving test and immediately got a vacation job with Ferguson's, a firm on the Royal Mile, which produced that delicious sweetmeat, Edinburgh Rock. Their regular van driver was taking a two-week Christmas holiday, and I was appointed as his substitute, surprisingly as I was a totally inexperienced driver. I had to load the van each morning with packing cases of Ferguson's Rock, and distribute these to sweetshops around the city. There was one delivery to a shop in Princes Street that was particularly nerve-racking. I distinguished myself by failing to lock the Ferguson firm doors when I closed shop for the New Year holiday. To my great relief this failing was undiscovered and no crates of rock were stolen.

For several summers in the mid-50s, I was one of a team of student 'beaters' employed on a daily basis to tramp over the heather-clad slopes in the Grampian hills near Aberfeldy, flag-waving and shouting in demented chorus in order to rouse the grouse and drive them towards the butts where wealthy elite, among them J. Arthur Rank and his party, were installed to shoot the birds as they flew overhead. Five long exhausting beats were accomplished in the course of a day for £1 remuneration, enough to finance a liberal beer-quaffing binge at the end of the week.

This grouse beating was all organised by my school friend Fergus Craik, the one since renowned for his life-long work on memory, whose father, the bank manager in Aberfeldy, had all the vital local contacts. Fergus and I were subsequently taken on by Donald Macdonald, ex merchant seaman from the Isle of Harris, who ran a grocery van from his base in the village of Foss to the hydroelectric scheme that was then under construction at Calvine on the River

Garry, some miles north of Blair Atholl. The labour force on this scheme was a rough lot, many Irish and Poles, who were earning good money. Fergus and I would load the van at Aberfeldy in the morning to meet their diverse requirements, then drive back over the hills by the village of Dull (proudly 'paired with Boring, Oregon, USA'), past Loch Tummel and on to Calvine, where we set up our roadside stall on a piece of land prepared for this purpose. We would stay till all the perishable goods were sold, then drive back to Foss in the evening, taking turns at the wheel. There we would sit with Donald making up the books for the day, income and expenditure, till he would announce in his dulcet Hebridean tones that 'sufficient unto the day is the evil thereof', and we would gladly take our rest.

In the summer of 1957, I found myself in Manitoba, working on the Trans-Canada pipeline, which was then being laid by the Bechtel Canada Co. some 20 miles south of Winnipeg. This pipeline was over a metre in diameter. A huge machine first scraped the rust off with a hideous screeching noise; this was followed by another huge machine that coated the pipeline with black bituminous paint; my job was to follow close behind this machine with a bucket of boiling tar and a brush, and fill in the seams between different sections of the pipe, the parts that the machine couldn't reach. It was exceedingly hot, the air was full of rust and the fumes of tar, and the day was very long; but the pay was good, and I earned enough in several weeks to allow me to hitch-hike the length and breadth of Canada. One of my rides was with a rich American who offered me a seat in

I hitch a ride by plane in Canada, 1957

his light aircraft for the next stage of my journey. To hitch a ride in an aircraft was an opportunity not to be missed and I readily accepted, although it took me quite a bit off my planned route!

RNVR

While a student at Edinburgh University, I joined the Royal Naval Volunteer Reserve (RNVR), which involved two weeks' training every summer. The first such period was in the summer of 1954, on the aircraft carrier *HMS Implacable*, anchored in Portsmouth (a ship that was decommissioned and sold for scrap one year later). On this, I learnt to sleep in a hammock and to drink a regulation tot of rum at 11 every morning! The following year, it was a cruise right round the British Isles in the minesweeper HMS Killicrankie, that sailed out of Granton harbour; and the year after that a cruise in the same old minesweeper across the North sea to Lubeck and then Bornholm Island.

Trinity College Cambridge

My Tutor when I finally arrived in Cambridge was Mark Pryor, who found an ingenious loophole in the University Ordinances that enabled me to count my second year in Cambridge both for the BA degree and as the first year of a PhD. This loophole, needless to

George Batchelor in 1959; photo by Chia Shun Yih

say, was swiftly closed by the Old Schools. My research supervisor was George Batchelor, world authority on the problem of turbulence. At that time the dynamics of electrically conducting fluids was in vogue, so I opted to work at the interface between these two areas of research, on the problem of magnetohydrodynamic turbulence, a topic of great relevance both in astrophysics and in the development of plasma containment devices like the tokamak. Batchelor was a model supervisor; I would give him screeds of immature hand-written work, which he would return to me the next day with copious marginal comments and criticisms. Just two years earlier, he had founded the *Journal of Fluid Mechanics*, one of CUP's most successful journals, and he brought a scrupulous editorial acuity to bear on any writings that were put before him. In spite of a rather austere temperament, he inspired great loyalty and affection in his students; for me, it was an intensely formative experience.

In those days, as a research student, I owned a 1937 MG sports car, which I habitually parked in Garret Hostel Lane, under the revolving spikes behind Bishop's Hostel. These spikes provided a convenient entry point to the College after midnight, when entry via the Great Gate would disturb the night Porters and lead to tutorial sanctions the following day. Climbing into College was the accepted price to pay for an evening on the town! DNA analysis of these spikes might provide fascinating evidence of past nocturnal activity.

With my sister Lindesay about to set off from Inveresk for Cambridge in my 1937 MG

Les Houches 1959

In 1959, I was a student participant, together with Donald Lynden-Bell at the *Ecole d'Eté de Physique Théorique* in the village of Les Houches in the Haute Savoie. We drove there in the MG, with Donald as navigator. That year the course was on the theory of neutral

The classroom at Les Houches 1959: Donald Lynden-Bell and I are in the front row, extreme right.

and ionized gases, and we had a brilliant series of lectures. Les Houches provides a spectacular setting for creative thinking, and I have returned there at every opportunity. The photo here shows Leon Van Hove (subsequently Head of the Theory Division at CERN) lecturing to the 1959 class.

Donald, an experienced mountaineer (and later an eminent Astronomer), induced me to climb the Aiguille du Bionnassay with him one weekend during the course. We reached the ridge to the summit at about 9 a.m. on the Sunday morning, already too late because the snow was melting and we sank knee-deep with every step. There we were, Donald and I, stuck on the ridge to the summit of Bionnassay, sinking to our knees in the deep snow that was melting as the sun rose,

couldn't go up, even worse couldn't go down. So we dug a hole in the shelter of a rock and settled down for 12 hours, pummelling each other all day to keep warm till the snow froze again at nightfall, and we were able to make our way down the treacherous slopes of the mountain. We descended at midnight under a full moon. During that unforgettable day, Donald taught me the laws of thermodynamics in a way I had never previously understood.

Driving home through the Jura with Donald at the end of the course, we hit a cow which stampeded across the road right in front of us. The farmer at the roadside had tried to restrain the cow, but to no avail. The MG, 'decapotable' with its sharp low bonnet, lifted the cow broadside-on, and it rolled over us, smashing headlights and wind-screen on the way, and bringing us rapidly to rest. The farmer, as alarmed as we were, took us into the farmhouse, and plied us with eau-de-vie to calm us down! We then continued our journey, thus happily fortified. I was even happier when I learnt that in France, it is the cow's insurance that covers such accidents, so that eventually my MG was repaired without cost to me.

In 1960, as previously related, I attended the Tenth International Congress of Applied Mechanics in Stresa in Northern Italy. I was immediately a convert to this series of Congresses, which bring together every four years a wonderful range of experts and personalities in the field of (now) *Theoretical and* Applied Mechanics.

Assistant Lectureship, and Fellowship at Trinity 1961

In 1961, I was very lucky to be appointed to a Cambridge University Assistant Lectureship, and then to a Teaching Fellowship at Trinity College. I gave my first course of 16 lectures on *Oscillations and Waves* during the Easter Term, April-June 1961. Martin Rees, former Master of Trinity, tells me that he attended this course as a first-year student; he was therefore a member of the large, noisy class into which I walked for my first lecture trembling with nervous anticipation. Without a word, I wrote on the blackboard a quotation from the poet Swinburne: *"And behold . . . the waves be upon you at last".* There was a reassuring cheer and applause and stamping of feet from the class; the ice was broken, my nerves recovered, and I was able to proceed, noteless in the style of Kemmer, and with no further difficulty. My career as a University Lecturer was launched.

CHAPTER 3

The 60s: Knotted Vortices

The title of this chapter relates to my 1969 discovery of a quadratic invariant of the very classical equations of fluid dynamics that date back to Euler (1756), a topological invariant to which I gave the name 'helicity'. Helicity represents the degree of linkage and/or knottedness of the vortex lines of any fluid motion. It was an exciting discovery, because the only quadratic invariant that had been previously known was the kinetic energy of the flow, invariant only insofar as viscous dissipation of this energy is negligible. Helicity is different from energy in that it is nonzero only if the flow lacks mirror symmetry (like a right hand whose mirror image is a left hand that cannot be moved into coincidence with itself; in this sense, either hand 'lacks mirror symmetry'). To find a new invariant of the Euler equations that had remained concealed for 200 years was amazing. I remember giving a lecture with the title *"The degree of knottedness of tangled vortex lines"* at the British Theoretical Mechanics Colloquium (BTMC) in Oxford in 1968. On the way into the lecture room, James Lighthill asked somewhat incredulously *"Have you really found a measure of knottedness of vortex lines?"*, to which I could only answer *"Yes, wait and see"*. So even Lighthill was surprised by the result.

Tying the Proverbial Knot 1960

Our wedding, December 1960

But first, let me take some important events in chronological order. In 1960, I managed to induce Linty to come down from Scotland for the May Ball, which was of course in June. This was a good move, because we were married before the year was out, and our first child, Fergus, was born nine months and a day later. Edge congratulated us on *attaining the minimum within epsilon* (a mathematician's way of saying "you cut it mighty fine")! For the first six months of married life, we lived in a caravan off the Hills Road, next door to Hammonds' Auction Sale-room, where we bought the furniture for our first home in Chedworth Street, in the so-called 'favoured Newnham area'. We purchased the house from Jack Hamson, Fellow of Trinity, for the princely sum of £2,200 (about £61,000 at 2024 prices).

Marseille 1961

One month before that happy event, I was seriously involved in the legendary *Colloque International sur la Mécanique de la Turbulence*, of which the principal organisers were my research mentor, George Batchelor,

Caravan life, Jan-June 1961

Alexandre Favre of Marseille, where the conference was held, and Leslie Kovasznay, then at Johns Hopkins University, Baltimore. My thesis work on *Magnetohydrodynamic Turbulence* was well under way, and I was in effect drafted in to lecture on this topic, a signal honour given the panoply of famous scientists that were attracted to the meeting. These included G.I. (Sir Geoffrey) Taylor, Theodore von Karman, and a group of Soviet researchers under the legendary A.N. Kolmogorov. [I remember the excursion to the beach at Cassis when Kolmogorov swam far off towards the horizon, much to the alarm of the western organisers; but his own colleagues reassured us that this was his normal behaviour, not to worry, and he would soon return.]

I have written at length about the scientific lectures and discussions at the meeting, and it would be superfluous to expatiate further here. Suffice it to say that this Colloquium not only broke new ground in turbulence research, but also highlighted areas, for example the phenomenon of 'intermittency', that would dominate research on turbulence for the decades that followed.

Department of Applied Mathematics and Theoretical Physics (DAMTP)

Batchelor had established the Department of Applied Mathematics and Theoretical Physics (DAMTP) in 1959, and was appointed its 'Head of Department' for the five-year period to 1964. There were just two Professors in the Department in 1959, Paul Dirac (1902--1984), Lucasian Professor of Mathematics (the Chair that had been held by Isaac Newton in years long gone by), and Fred Hoyle (1915--2001), Plumian Professor of Astronomy. Dirac was aloof and disinterested in Department affairs, preferring to work from his rooms in St John's College. I attended his course of lectures on Quantum Mechanics, in which his technique was to hold his famous book in one hand, and transcribe its content with the other to the blackboard, which, in those days, was the sole visual aid. I also attended Hoyle's course of lectures, or at least the start of it, on stellar structure, but I found his style to be so chaotic as to be (for me) totally incomprehensible. The best lecturer at that time on the Applied side was Ian Proudman, who was soon to move to Colchester to help found the University of Essex, following his affair with Batchelor's secretary and the resulting break-up of his marriage. I believe that my appointment to a teaching Fellowship in Trinity College in 1961 was in anticipation of the vacancy caused by Ian's departure.

Our second son, Peter, was born just before Christmas 1962, the beginning of the most severe winter in living memory, when the Cam was frozen till March 1963.

Skating on the river Cam, January 1963

I take Peter for a walk in the Fellows' garden, Trinity College, January 1963

In 1962 we still occupied a set of rooms in Free School Lane. It was in October 1962 that Stephen Hawking arrived in the Department, to undertake research towards the PhD degree under the supervision of Dennis Sciama. I remember him then as an eccentric, even droll, young man with a velvet jacket and a floppy bow tie; he was prominent, perhaps deliberately so, among the new arrivals. It was somewhat later, when Stephen fell down the stairs in Free School Lane, that the severity of his physical condition came to be recognised. He was diagnosed with a form of motor neurone disease within a year of his arrival in Cambridge. From then on, it was a tale of incredible courage and dazzling scientific genius.

In 1962, Cambridge University Press (CUP) announced its intention to vacate its 19th century printing shop and warehouse behind the very classical Pitt Building on Trumpington Street. Batchelor immediately made a bid to convert the machine shop for DAMTP, which was already sore pressed for space in Free School Lane. He asked me to liaise with Estate Management on all aspects of the conversion, which went ahead throughout most of the 60s and 70s. The actual move into the new premises began in 1964 in time for the start of the Michaelmas Term that year.

One of my research students during that period was Jüri Toomre from USA, who has since enjoyed a distinguished career in Astrophysics at the University of Boulder, Colorado. He and his wife were equally prominent in the Department, as they owned a magnificent Afghan Hound, a regular visitor to the Department! Jüri was a keen photographer, and helped to design a photographic darkroom that served the Department well for all the photographic developing and printing that had to be done in those days.

I remember Jüri coming into my office on 22 November 1963, his face as white as a sheet, to tell me that President Kennedy had been assassinated, a shocking event in world history. We had a more light-hearted exchange on another occasion when Jüri asked me if there was to be a laboratory in the new premises. He pronounced **lab**oratory in the American way, with emphasis on the first syllable, and I misheard his question as "Will there be a lavatory in the department", which I thought to be a rather strange enquiry. However, I answered "Oh, yes, one on each floor, with the Ladies one on the top floor". Now it was Jüri's turn to look puzzled!

Zakopane 1963

George Batchelor had been active for some years in promoting (in collaboration with Wladek Fiszdon and Ryczard Hercynski) the biennial fluid mechanics conferences that were held in Poland from 1959 on, during the cold-war years; these meetings provided a valued opportunity for scientists from the old USSR, who were able to travel to Poland although not further west, to interact with scientists from Europe and USA, who were also able to travel to Poland (if armed with appropriate visas). The meeting in 1963 was to be held in Zakopane in the high Tatra mountains on Poland's southern border with Czechoslovakia (as it then was). I was then writing my paper on corner eddies and Batchelor suggested that I should lecture on this work at the meeting. He went further and offered to drive me there in his six-seater Ford Zodiac, together with his wife Wilma and his three daughters, Adrienne, Claire and Bryony. This was an offer that I clearly couldn't refuse. We drove to Harwich and crossed by sea to Hook of Holland. At that point. Wilma noticed that George was still wearing his bedroom slippers, having forgotten to change them before departure from Cambridge. It is symptomatic of the straitened circumstances of those times that the possibility of buying a new pair of shoes simply did not arise. Instead George phoned Brooke Benjamin, who was later going to travel to Poland by train, and arranged for him to find George's shoes and bring them with him to Zakopane. This he duly did, and the crisis was satisfactorily resolved in time for George's own lecture at the Conference, although not for the several days of driving across the Netherlands, Germany, Czechoslovakia, and Poland itself. There were few roadsigns in Czechoslovakia, and little traffic and we had to rely on a compass and GKB's infallible instinct to keep us on our planned route via Prague. On the return journey, we headed for Bratislava and Vienna, where the car broke down with what amounted to a sigh of relief that we were by then back through the iron curtain.

I was intrigued when we passed through the town of Katovice to note that it was pronounced locally to rhyme with Nietzsche. I am always tempted to compose a limerick when such rhymes present themselves, and this was no exception. But it was some years later that I composed this one, having an academic intention, and yet an unavoidably scurrilous flavour:

George Batchelor and family en route to Zakopane

A lady from old Katovice
Wanted to work under Nietzsche.
Show me, he said, how you function in bed
Then I'll be happy to teach'ye.

More sombrely, we visited Oświęcim (i.e. Auschwitz), on our way to Krakow. Oh my God, what a dreadful place it was still then, and no doubt still is now. The smell of death still lingered there in a truly awful setting. I still remember the sense of horror that overcame me in viewing the relics of the atrocious crimes against humanity that were committed there.

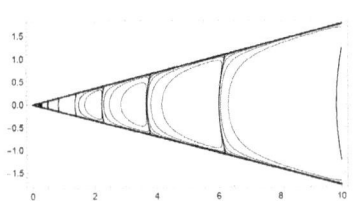

Corner eddies: I presented the theoretical derivation at the 1963 Zakopane meeting

In Zakopane, I duly gave my lecture on corner eddies, which must have impressed Milton Van Dyke, who later invited me to spend a semester in Stanford, California, a wonderful opportunity that I gladly accepted, as soon as my first sabbatical permitted.

The Zakopane meeting was greatly enlivened by the presence of the Soviet participants, even though they were closely guarded by KGB minders. One evening at a social gathering, one of the young Russians got up and sang a beautiful soulful Russian song, but was silenced by the minders because of its nostalgic undertones. A more amusing episode occurred at breakfast one day when three of us westerners were seated at a table for four; a Russian arrived and was welcomed at the fourth place. There was a pot of jam on the table intended for all four, but he assumed it was his alone, took it, and cheerfully consumed the lot. We could only look on, in silent wonder!

The Exodus of Fred Hoyle

Back in Cambridge, a drama of another sort erupted in 1964, when Batchelor's first term of office as Head of Department came to an end. He wished to continue for a second five-year term, but this was contested by Fred Hoyle, who disliked Batchelor's style of micro-management. A Departmental vote on the matter came out clearly in favour of Batchelor, whereupon Hoyle walked out in high dudgeon, never to be seen in DAMTP again. He proceeded to set up the Institute of Astronomy on Madingley Road, aided by Ray Lyttleton who left DAMTP at the same time. Batchelor made a bargain with the University that, when Lyttleton eventually retired (as he would have to do by 1978 at latest), his post would then revert to DAMTP. [This was a far-sighted bargain, much to my personal benefit, because

I was elected in 1980 as the first holder of the resulting 'Chair of Mathematical Physics (1978)'.]

Move to the Press site

In 1964, the fluid mechanics and astrophysics/cosmology groups in DAMTP moved to the Press site at the new Silver Street address. Estate Management warned us that the building was technically condemned, and would last only 5 years; in the event, it served us well for 35 years, and when we finally moved to new accommodation in Wilberforce Road, other smaller Departments were quick to move into the space released in this prime position. Brooke Benjamin became Director of the new Fluid Dynamics laboratory in the basement; I recall how expert he was at building intricate control mechanisms using a simple Meccano set! The Common Room was the 'beating heart' of the Department; here we would all meet for morning coffee, when problems could be informally discussed around formica-topped coffee tables on which we could sketch out ideas with the felt-tip pens clipped on the side. Many key ideas were conceived in this congenial setting. And the fluid dynamics seminars, 4.30 on Fridays, provided an opportunity for more formal but equally vigorous discussion, often adjourned with the speaker to the Anchor (the riverside pub by the Silver Street bridge), and continuing till 7 or later in the evening.

In 1964 also, Linty and I, together with our two young sons, Fergus and Peter, moved house in Cambridge from

Invitation to a party; four of our Churchill College students, c 1967

Chedworth Street to 67 Barton Road, our home for the next 13 years; this was a large detached seven-bedroom house with a garden of one-third of an acre, ideal for a young family. For several years, we let three rooms on the top floor to undergraduates from Churchill College, which was still constructing its student accommodation, or to students from the neighbouring language school (now the Kaplan School).

It also had a basement where for several years I brewed elderflower champagne. One year, I made the mistake of bottling this in tightly screwed screw-top bottles, which exploded when I was absent overseas at a conference. Linty investigated and found that a chain-reaction had occurred --- 12 bottles had exploded and a thin mist of minute glass particles was slowly settling to the floor. Linty hastily retreated and the cellar remained out of bounds until I returned to face the damage.

Stanford 1965

The Edsel Citation convertible at Alvarado Row

As soon as my sabbatical came up in March 1965, we set off for California, first by the ocean liner 'SS United States' to New York. This was quite an experience, although I can't say it was a pleasant one, as the ship was trying to beat its own speed record for the voyage and was doing so in stormy conditions without using its stabilisers. It was a relief when the Statue of Liberty hove in sight, and we were glad to be grounded among the stevedores, while they unloaded all the cargo from the ship. We then flew in relative luxury to San Francisco, and on to Alvarado Row on the Stanford campus, which was to be our home for the next 4 months. As soon as my first pay-check as an Assistant Professor arrived, we bought for the princely sum of $130 a second-hand Ford Edsel Citation convertible, which served us well for all our American peregrinations.

In Stanford, I gave a graduate course of lectures on *Turbulence*. One student, Fazle Hussain, excelled in a class of about 20; he has since become a world authority in the subject. The course led me to study the action of a strong shearing wind on atmospheric turbulence; I was invited to an International Colloquium on *"Atmospheric Turbulence and Radio Wave Propagation"* in Moscow in June that year. This was the ideal opportunity to present this work, so I made the journey, abandoning Linty with Fergus and Peter for the duration --- it seems unconscionable to me now that I could do this, but Linty, as always, gave me every encouragement. The Moscow meeting, organised by A.M.Obukhov, was brilliant. My lecture was simultaneously translated and delivered in Russian by Georgi Golitsyn, scion of a princely dynasty. I didn't realise it at the time, but my lecture (published in the Proceedings of that Colloquium) contained the first description of what later came to be recognised as a 'transient instability', one that grows in linear manner and persists for a very long time, but is ultimately suppressed.

Fergus and Peter, Stanford 1965

Back in Stanford, we had a trip by car down to La Jolla, where I gave a seminar at IGPP (the Institute of Geophysics and Planetary

Physics). We met Sir Edward (Teddy) Bullard there one day, and walked the beach with him. Walter Munk, legendary oceanographer who died in 2019 at the age of 101, entertained us at his home with his charming wife Judith. John Elder took me on a memorable trip across the southern border to Tijuana, where we watched a brilliant game of Jai alai and went on to several very raucous night clubs; of that, the less said the better!

Visiting my Auntie Phoebe and her family in Ottawa, August 1965

In August we drove back across the States via Nevada (no speed limit), Utah (the salt flats and Salt Lake City), then somehow up through Wyoming to camp among the bears in Yellowstone National Park, then on through North Dakota and across the border to Winnipeg, to stay for some days with Linty's twin brother Bobby and his wife Marlene. From there, we made our way to Sault Ste Marie, and on to Ottawa to stay with my Auntie Phoebe, her genial husband John Hutchinson, and their six children, Ian, Jamie, Kennie, David, Eleanor and Peter -- a very lively family! They were very impressed with our Edsel!

Johns Hopkins University

From there, we drove down to New York, where we did the sights for a few days. Linty flew home with the boys in September, and I drove on down to Baltimore, to spend a term at Johns Hopkins University at the invitation of Stan Corrsin, doyen of turbulent diffusion theory. He held a weekly lunchtime seminar, at which we struggled to understand Kraichnan's emerging 'Lagrangian history direct interaction approximation' for the 'closure problem' of turbulence, a struggle that left me exhausted and unenlightened in equal measure. There was a certain tension within Corrsin's department caused by the presence of Clifford Truesdell, who scorned any approach that was not strictly rational and analytic, and put him at loggerheads with most other fluid dynamicists of that period. Clifford did invite me to dinner at his home, and placed me, as I learnt was his custom, opposite a life-sized nude portrait of his wife who sat at the same table; I could not do other than admire the portrait!

With George Batchelor and *JFM*

Owen Phillips, known especially for his seminal work on wind-wave interaction, was already settled in Corrsin's department, having spent a brief spell at DAMTP in Cambridge. He invited me to a Thanksgiving party at his home, where his wife Mearle served traditional turkey, with pumpkin and all the trimmings. I returned to Cambridge in time for Christmas 1965. George Batchelor immediately roped me in to serve as co-Editor with him of the *Journal of Fluid Mechanics*, a massive burden of responsibility that I held for the next 18 years.

Our first daughter, Hester, was born in April 1966. The name Hester had come to our attention somewhat earlier when Lord Adrian, who had been Master of Trinity when I was elected a Fellow, had invited Linty and me to a dinner party in the Master's Lodge. Drinks were served beforehand in the magnificent drawing room overlooking Great Court. We were there when the Nobel laureate Laurence Bragg arrived; he strode into the drawing room with great panache, and embraced the hostess with the greeting "Ah, Hester", for she was indeed Hester, Lady Adrian. So our Hester is in effect named after Hester Adrian.

Our second daughter, Penelope, was born in September 1967; so now there were four, Fergus, Peter, Hester and Penelope. One of our lodgers at Barton Road around this time was Antonio Tablas from Lanestosa in the Basque region of Spain. He arrived speaking no English, but was learning fast with the chatter of our children all around him. Once, we had a full-day family outing, and we left Antonio with instructions to feed the cat while we were away. When we arrived home, he greeted us with a huge smile and said proudly *"I have eaten the cat"*, at which our children burst into tears and were inconsolable until the language difficulty had been resolved, and (like Schrödinger's cat) the cat's living existence could be established.

So now there were four: from left, Hester, Peter, Fergus and Penelope, December 1967

Strawberry Hill 1967

By July 1967, My parents had settled well in at Strawberry Hill, near Irish Town in the Blue Mountains of Jamaica, and I flew out with our two boys to spend two weeks with them there. Strawberry Hill is located about 3000 ft above Kingston, so free from the suffocating heat and pollution of the city. It was then an old colonial-style house with a number of cottages equipped for visiting guests. The grounds extend over several acres, with banana groves and lush tropical foliage in abundance. My parents had fallen in love with the place when they had stayed there earlier the previous year; "This is my picture of Paradise" enthused my mother to the owner, Mr Da Costa, who responded "You can buy it if you like; it is for sale". Within a week, the purchase was completed, my mother being very impulsive! They moved there from the UK in the summer of 1966, taking my 89-year old Granny Fleming with them.

My grandmother and mother in the garden at Strawberry Hill, 1967

Peter and Fergus join me for a swim in the pool at Strawberry Hill, 1967

Helicity 1969

For several years in Cambridge, I had been giving a Part III (graduate) course on Magnetohydrodynamics, a subject that embraces both Fluid Dynamics and Electromagnetism. I had 'inherited' this course from Arthur Shercliff, when he left Cambridge in 1964 to take up the Chair of Engineering Science at the new University of Warwick. Each year, I included an account of a result that had been discovered by the Astronomer L.Woltjer in 1956, namely that, under certain conditions, the space integral of the scalar product of a magnetic field and its vector potential is invariant in a moving fluid, provided the fluid is perfectly conducting. This result was quite easily proved mathematically, but its physical meaning was not evident. I found it awkward to present this result in lectures without being able to provide a physical interpretation, and I struggled with this for several years. Then, in 1968, the light dawned: I realised that Woltjer's result held because, in a perfectly conducting fluid, the magnetic lines of force are transported with the fluid just like elastic fibres. This means that the topology of the field is invariant; in particular any knots or linkages of magnetic field lines must be conserved, however complex the fluid motion may be. Woltjer's invariant was therefore a measure of this conserved topology.

But then a similar result must hold in an ideal fluid (zero viscosity), for which, according to the 19th century laws of Helmholtz and Kelvin, vortex lines are similarly transported with fluid motion governed by the very classical Euler equations. The quantity analogous to Woltjer's invariant is the space-integral of the scalar product of velocity and vorticity, and I proved that this quantity is indeed an invariant of the Euler equations. I deliberated for some weeks on a suitable term for this new invariant, and eventually settled on the term *'helicity'* in my 1969 paper *"the degree of knottedness of tangled vortex lines"*, a term which rapidly became established in the literature of fluid dynamics: if I browse google scholar now for *'helicity in fluid mechanics'*, I get more than 22,000 hits!

But this was by no means the end of the story. It happened that, nine years later in 1978, I applied (unsuccessfully) for a vacant Professorship at the University of Grenoble. The application was considered in Paris by the French National Committee for such appointments. A member of this Committee, Jean-Jacques Moreau, of the University of Montpellier, having read my cv, wrote to me expressing interest in my work, and enclosing a copy of his 1961 paper *"Constantes d'un îlot tourbillonnaire en fluid parfait barotrope"*, published in the *Comptes Rendus* of the French Academy. Sure enough, Moreau had discovered the invariance of helicity, although he didn't call it this, nor did he recognise the link with the earlier work of Woltjer. Nevertheless, I had been scooped. I cited the paper as soon as I knew of it, in my 1981 paper *"Some developments in the theory of turbulence"*, which then helped to give it the publicity that it deserved; I believe this was its first citation. Somewhat later, I met Jean-Jacques Moreau at a Conference in Grenoble, where he was kind enough to say that I had recognised the importance of the result and knew what to do with it. And that was true, because it turned out to be important for both dynamo theory and for the general dynamics of turbulence.

Duplicate scientific discoveries: it happens quite often!

There are many similar examples of discoveries in science that are made quite independently at widely separated intervals. One came to my attention soon after I became an Editor of *JFM*. This related to Robert Betchov's paper *"On the curvature and torsion of an isolated vortex filament"*, which had been published in *JFM* in 1965. Some years later, I received a letter from Massimo Germano of the University of Turin, who drew my attention to a paper by L.S. Da Rios, *"Sul moto di un liquido indefinito con un fileto vorticoso di forma qualunque"*, published in 1906 in the Italian journal *Rend. Circ. Mat. Palermo*; it appeared that Da Rios had obtained precisely the same results as obtained quite independently 60 years later by Betchov. I cited the Da Rios paper in my 1984 paper *"Simple topological aspects of turbulent vorticity dynamics"*; I do not know of any earlier citation of this work.

A further example: in 1916, the eminent Lord Rayleigh published a result concerning the instability of flow with cylindrical symmetry; in simple terms such a flow can be unstable only if the circulation decreases outwards with radius, a condition that came to be known as 'Rayleigh's criterion' for the instability of circulating flow. But this result was actually known to James Clerk Maxwell 50 years earlier. For in 1866, Maxwell was 'moderator' for the Smith's Prize examination at Cambridge University. One of the problems that he set concerns just this type of flow. The sting in the tail of this question was in the final sentence: *"Hence show that the motion of a fluid in a circular whirlpool will be stable or unstable according as the areas described by particles in equal times increase or diminish from centre to circumference"*. This is precisely Rayleigh's criterion, although expressed rather differently.

Student unrest 1969

Rab Butler ('the best Prime Minister we never had') had succeeded Lord Adrian as Master of Trinity in 1965. It is believed that Harold Wilson recommended his appointment to the Queen, in his desire to remove Rab from the opposition benches in the House of Commons. The Prince of Wales, now King Charles III, studied at Trinity 1967--1970, while Butler acted as his 'mentor and counsellor'.

Student unrest was growing worldwide in 1968/9, and Trinity was not immune from protests against academic authority. I had been drawn into College politics when I succeeded Jack Hamson as Senior Treasurer of the College [Students'] Union, the students believing that I would be more sympathetic to their demands. In this, they were only partially correct; but at least I lent them a more sympathetic ear. In this capacity, I was elected to the College Council in 1968.

Things came a head in February 1969, when an extraordinary meeting was held in the large Elizabethan Hall of the College, to address a string of grievances that had been put to the College Council in the preceding weeks. All senior and junior members were invited to this meeting, and the Hall was packed. Members of the College Council were at the High Table end, and Rab, as Master, was in the Chair. There has never been another such meeting to my knowledge, either before or since. It is historically significant that the *Admission of Women* to the College was for the first time debated at this meeting, although a further nine years would elapse before women undergraduates were actually admitted for the first time. I should note that there were at that time five Tutors in the College, each having a 'tutorial side' of up to 160 *"men in statu pupillari"*.

Each member of Council had been assigned one of the grievances raised by the students, to address in turn. The Master had asked me to address the question of 'Gate Hours', i.e. the nocturnal hours when the undergraduates had to be inside the College gates, a constraint that they naturally resented. Many other issues had been debated by the time I was called upon to speak, and we were all rather jaded. I had prepared both a serious speech and a humorous speech laced with irony, and at the last moment I chose to use the latter in order to lighten the atmosphere, which it undoubtedly did. This was in response to the proposal from the undergraduates *"That provided a practical solution is possible, Gate Hours should be uncontrolled".*

Forty years later, I was amazed to receive a letter from Clem McCalla, a mathematics student from Jamaica who had been President of the College Union in 1969. He wrote: *"Attached find a text of the speech that you gave to the College Meeting on 25 February 1969. If you recall, I was so enthralled by your speech that I asked you for the text. On returning to my home in the US, I found a hand-written copy in my own handwriting in my records. As Xerox machines were not widely available in those days, it appears that I had transcribed a copy of your text of your speech (whether type-written or hand-written) into my own handwriting. Making use of this modern-day technology, I have gone one step further to transcribe the text into a Microsoft Word document."* I had not retained a copy of my speech, so was very glad to have this copy that Clem had so carefully retained in his files (see below). I am gratified to note that, in my final remarks, I anticipated the face-recognition technology that was actually realised some 50 years later!

My speech to the College meeting, 25 February 1969

Master, I shall talk only on the proposal relating to Gate Hours. I am placed in the fortunate position of being called upon to propose a motion to which no sane person could possibly object.

Our rules regarding gate restrictions are totally archaic. Undergraduates at the moment are required to return to College every night at the intolerably early hour of 2 a.m. If they wish to stay out later that that, they must go through the humiliating and time-consuming process of signing their names as they come in and they are allowed to do this only 5 times each term without special permission.

One of the Tutors was kind enough to show me his statistics on this. He has a side of about 130 men excluding BA's, and it seems that no less than 2 of them actually used the maximum number of late leaves during the Michaelmas term and the nocturnal activities of these two were no doubt gravely inhibited as a result. I have no doubt that this side was not untypical and that it gives a fair indication of the severe distress that is undoubtedly caused by the present situation.

Then Master, there is the special problem of the men who live in Whewell's Court. As if these men are not adequately discriminated against already, they must suffer the additional inconvenience of being unable to escape from the court for a breath of fresh air after 12 midnight. They can get in until 2 a.m. but once in, there is no escape. There a man is imprisoned for the whole night long – he can't slip out to visit a friend, he can't even get out for a quick supervision. And yet we talk of freedom.

Master, there can be no question that these rules are unfair, arbitrary, illogical, unnecessary, invidious, vexatious, odious, offensive, obnoxious – and very unpopular. I hope that they can be abolished.

The practical difficulties that are referred to in the Committee report are, I am sure, surmountable. An electronic gate is the idea that we favour. The gate would be fitted with an electronic eye which would recognise members of the College but no one else. Ideally, the gate would open automatically as the College member approached, like these automatic doors in American airports, and would close silently behind him. Master, this is an idea for the future and I hope that this is the direction in which we are all looking.

Mean-field electrodynamics 1966/70

Dynamo theory is a branch of magnetohydrodynamics that seeks to explain the generation of the observed magnetic fields in cosmic bodies --- planets, stars and galaxies. One of my research students, Glyn Roberts, wrote an important paper arising out of his 1969 Cambridge PhD. In this paper, published in the Philosophical Transactions of the Royal Society in 1970, he established the possibility of dynamo action (i.e. exponential growth of a magnetic field starting from an infinitesimal level) by a space-periodic flow. I had urged Glyn to generalise his work to a turbulent velocity field, which in my view would have much wider generality, but Glyn did not take up this suggestion, and he moved to Newcastle in 1968 to join the research group of Paul Roberts (no relation!).

I thereupon myself carried out the generalisation to turbulence that I had advocated, and sent a draft paper to Glyn in November 1968. Glyn responded immediately that he had attended a conference in Madrid, at which Fritz Krause from the Astrophysics Institute in Potsdam (in East Germany, then firmly behind the iron curtain) had presented a theory very similar to mine, and that this theory had been developed in a series of papers, published in German in the Journal *Zeitschrift für Naturforschung*, since 1966. These papers were hardly known to western scientists at the time. I read them with the help of Bruno Renner, a German researcher in DAMTP, and realised that once again I had been scooped. I rewrote my paper, giving full credit to the Potsdam group, with opening paragraph: "*A theory that is likely to be of the greatest significance in geomagnetism and in cosmical electrodynamics has been developed recently by Steenbeck, Krause & Rädler (1966), Steenbeck & Kranse (1966, 1967), Rädler (1968) and Krause (1968). The theory is concerned with the effect of a turbulent velocity field on a magnetic field distribution in an electrically conducting fluid, it being supposed that there is no external source of magnetic field, the only source being the electric current distribution within the fluid itself."* My paper was delivered at the *Boeing Symposium on Turbulence* (Seattle, June 1969), and published in *JFM* in 1970. There were no citations in the western scientific literature of the Potsdam papers, so far as I know, before 1970. They were translated into English by Paul Roberts and Michael Stix in an NCAR report published in 1971. My prediction that the theory, which came to be known as 'mean-field electrodynamics', would *"be of the greatest significance"* has been amply borne out by subsequent developments in this major branch of astrophysics.

E.N. (Gene) Parker also recognised the importance of the Steenbeck, Krause & Rädler [SKR] work, because he too published a paper in 1970 citing SKR, the first of a series of six papers published in quick succession in *Astrophysical Journal*, concerning various applications of the theory. 1970 was indeed a pivotal year for dynamo theory, which developed with a great burst of energy throughout the 1970s.

CHAPTER 4
The 70s: Dynamo Action

Mean-field electrodynamics took off in a big way in the 1970s, and I was very fortunate to be involved in this from the very beginning. The lack of mirror symmetry was quickly seen to be a crucial property for the spontaneous growth of magnetic field from an infinitesimal level in any sufficiently large body of rotating cosmic matter (e.g. fully ionised gas as in much of the Milky Way); and the large-scale magnetic field generated would exhibit similar lack of mirror symmetry in the form of magnetic knots and/or links. I like to summarise the situation with the rhyme:

> *Convection and diffusion*
> *By turb'lence with helicity*
> *Yields order from confusion*
> *In cosmic electricity.*

My second paper on the subject in 1970, "*Dynamo action associated with random inertial waves in a rotating conducting fluid*", incorporated the approach of mean-field theory into a 'dynamic' treatment, i.e. one involving solution of the Navier-Stokes equations coupled with the purely 'kinematic' magnetic induction equation obtained on the basis of Maxwell's equations. This paper showed that, in a rotating fluid, a flux of energy parallel to the rotation vector is enough to give background turbulence the 'lack of mirror symmetry' that is the key requirement for effective dynamo action. I followed this in

1972 with a further paper *"An approach to a dynamic theory of dynamo action in a rotating conducting fluid"*; in this, I invoked a random body force (which could be of thermal origin); the dynamo growth of a large-scale magnetic field then saturates at a level that I was able to determine. I believe this was the first paper to attempt a dynamically consistent dynamo theory, in which a 'quenching effect', which limits the dynamo growth of a magnetic field, was clearly identified.

NATO Advanced Study Institute 1972

I organised a NATO Advanced Study Institute in Cambridge in June 1972. Miraculously, Fritz Krause and Karl-Heinz Rädler from Potsdam were able to attend. We were lucky to have Sir Edward Bullard on the Organising Committee, and he spoke at the dinner held in the Hall of Trinity College, commenting particularly on the welcome presence of representatives from the German Democratic Republic (GDR, i.e. East Germany) at a NATO meeting; we hoped this would not cause trouble for Krause and Rädler. The meeting was historic in bringing the Potsdam work firmly to the attention of the western Astrophysics and Geophysics communities.

Potsdam 1972

Krause invited me to Potsdam shortly after this Conference, a visit that was quite an eye-opener for me. Here I met Günther Rüdiger, who has recently written a history of astronomy in Potsdam, which describes the complex political manœuvres that constrained the work of astrophysicists in the GDR during that period. This visit was a great privilege, as visits to science establishments in the GDR were quite difficult to arrange. I have a warm recollection of that visit, when I was able to have valuable scientific discussions with Krause and Rädler, and also with Rüdiger, who, as I recall with much pleasure, was my guide on a visit to the forest location where the famous 1945 Potsdam Declaration had been signed; but the Sanssouci Palace of Frederick the Great seemed (from a distance) to be in a sad state of disrepair.

I recall also the highly competitive games of table-tennis during the lunch breaks at the Astrophysical Observatory Potsdam (AOP); also a very productive mushroom-hunting expedition with Fritz Krause in the nearby meadows one day after work.

Karl-Heinz was also exceptionally kind during this visit inviting me to his home to meet his wife in a very friendly and welcoming atmosphere. His role in the AOP group gained prominence in the 1970s, culminating in the publication in 1980 of the Krause & Rädler monograph "*Mean Field Magnetohydrodynamics and Dynamo Theory*", which provided a systematic treatment of the mean-field approach, by then gaining widespread acceptance.

It is a sad fact that documents released after the fall of the Berlin Wall in 1990 revealed that Krause had been an informer of the STASI, presumably reporting on activities of his colleagues as well as of visitors to the Astrophysics Institute like myself. Rüdiger has recently confirmed to me that this revelation came as a great shock to the Astrophysics community in Potsdam, and led to the severing of the formerly close relationship between Krause and Rädler from that point on. It would appear that Krause had become entangled with the STASI to such an extent that he could not extricate himself without suffering severe consequences for himself and his family. His fall from grace in the 1990s was, in Günther's words, nothing short of a Greek tragedy. It is hard to comprehend the pressures that people like Krause in positions of authority must have laboured under, in the extremely oppressive DDR regime.

Białowieża 1973

I presented a review of my work on helicity and associated dynamo action at the biennial Polish Fluid Mechanics Symposium 1973, held that year at a Conference centre in the wonderful forest of Białowieża in the far East of Poland. At this memorable meeting, I met two great Russian scientists, Ya. B. Zel'dovich (1914--1987) and Olga Ladyzhenskaya (1922--2004). I gathered later that, on his return to Moscow, Zel'dovich communicated my result on the invariance of helicity to V. I. Arnol'd (1937--2010), who immediately developed the topic in characteristically rigorous manner (Arnol'd 1974, *The asymptotic Hopf invariant and its applications* [in Russian]; this was published in an inaccessible Proceedings of the *Summer School in Differential Equations*, Erevan, Armenian SSR Acad. Sci.) I met Arnol'd much later in Moscow in 1982; he gave me a copy of this 1974 paper with the comment that no-one in the West had read it (as was true beyond doubt)! It was much later republished in English translation (Arnol'd 1986, *Sel.Math.Sov.* 5, 327-345), and has since attracted the attention it always deserved.

College complications

In the early 70s, I became increasingly involved in College matters. The tutorial system had been overhauled in the late 60s, so that newly appointed Tutors would have on their 'sides' only 80 tutorial pupils, and would have no direct responsibility for admissions to the College, a responsibility that was passed to Directors of Studies in the various Tripos subjects. In 1970, already since 1961 a Director of Studies in Mathematics, I was appointed a Tutor, succeeding Theo Redpath, who was reluctantly relinquishing the post, having served the statutory 10 years as Tutor of 160 men *in statu pupillari*; I took on the A to Ms of this cohort. The N to Zs were taken over by Tony Weir, Fellow in Law, who occupied rooms in A staircase, Nevile's Court. I moved into the set immediately below his on the same staircase. Actually, Tony was on leave for the Michaelmas Term of 1970, so I had to look after the whole side of 160 for that term. I was astonished to find a remarkable proportion of Etonians among them. [One of them, during my first term as Tutor, distinguished himself by driving his car to Cambridge Railway Station to pick up a friend; in an excess of exuberance, he drove right through the plate-glass entrance to the station and onto Platform One, causing great consternation.]

And what a term it was in other ways! Early in the term, on a Monday morning, a distraught student came to me saying he was under investigation by the police for a murder that had been committed in Cambridge the previous Thursday evening. It seemed that the suspect had been seen wearing a yellow jacket just like the one that my student habitually wore. I asked him what he had been doing on that Thursday evening, and he in some embarrassment told me, as he had told the police, that he couldn't remember. I told him that he must wrack his brain, such as it was, and recall his exact movements before the police interviewed him more formally. Fortunately, the police found the true criminal within the next week, and the situation was thankfully resolved.

But then, a week or two later, another distraught first-year student embarking on a degree in English came to me early one morning. He had had a supervision with a female supervisor in Newnham College the previous evening, and had fallen madly in love with her at first sight. He had evidently protested his love to such an extent that the lady became quite alarmed and, he claimed, plied him with sedative and allowed him to spend the night on her sofa. This very immature 18-year-old arrived in my office in a highly emotional state begging me to advise how he could continue to live in a state of unrequited love. I did my best to calm him down, and arranged a change of supervisor without delay. Fortunately, his falling-in-love crisis dissipated as quickly as it had erupted, and I heard no more about it.

Les Houches 1973

In 1973, George Batchelor was invited to give a course of lectures on fluid mechanics at the Ecole d'Eté de Physique Théorique, at Les Houches in the Haute Savoie, where I had myself been a student back in 1959. George was, for some reason, unable to accept this invitation, and suggested that I should take his place, to which I happily agreed. We set off with our four children, now in a VW Dormobile, and spent six weeks in one of the log cabins of the Les Houches school, for the duration of the course. The students were predominantly French, but with a sprinkling of other nationalities. I gave my lectures in a mixture of French and English.

Hester, off to France with the VW, 1973

I was aware of a group of older-than-average 'students' in the class, among them, as it transpired, Pierre-Gilles de Gennes, Professeur at the Collège de France, and Etienne Guyon, Professeur at the Université de Paris-Sud. Both were already famous for their work on liquid crystals, but were minded to move towards basic fluid dynamics, which is why they had come to attend this course. There is no doubt that French research activity in fluid mechanics gained a huge boost from this change of direction of these giants of the subject; it has gained enormously in strength since that time. I like to think that my lectures, and those of Stephen Orszag who gave a parallel course on turbulence at the same summer school, may have helped in promoting the renewed vigour of research in fluid mechanics in France dating from about this time. Indeed, Etienne Guyon wrote in the Introduction to the 3rd Edition (2015) of his book *Physical Hydrodynamics* (with three co-authors) *"The teaching of John Hinch, who has accepted to write a "Forward" to this book, and Keith Moffatt who introduced fluid mechanics to a number of us in a famous summer Institute in Les Houches in 1973, have provided us with a number of fine tools suitable both for our research as well as for graduate and undergraduate teaching at the origin of this book"*. One could hardly wish for a more positive endorsement!

Trip to Lanestosa

After the end of the Les Houches course, we planned to visit the family of our former lodger Antonio in his home village of Lanestosa in the Basque region of Spain. We drove across the south of France and into Andorra, where we stopped for the night. Here Linty and Penelope, now 5 years old, had a narrow escape. They were about to cross the road when a lorry passed down the steep hill that runs through this tiny principality. About 200 yards behind it, a car came hurtling down, quite out of control. It rapidly caught up with the lorry and bounced of it three times before crashing into a bank at the side of the road. A shaken driver emerged. It transpired that his brakes had failed, and he was lucky that the lorry had been able to help to bring him to rest, albeit in a somewhat unconventional manner.

The next day, we continued our journey along the north coast of Spain, turning south at Laredo for the further 30 km or so to Lanestosa. The whole journey was far more arduous than we had anticipated, and we didn't arrive till about 2 in the morning. To our great surprise and delight, we found that a fiesta was in full swing, and was set to continue until at least 5 a.m. Antonio had rooms prepared for us, so Linty and I settled the children, and proceeded to drink brandy and to join in the fiesta celebrations for the rest of the night.

ICTP Trieste 1974

In 1974, I was invited to give a short course of lectures at the International Centre for Theoretical Physics, a few miles up the coast from Trieste in Italy. The founder of the Centre, Abdus Salam (Nobel Laureate 1979) was there at the time, and I was honoured to meet him. A lecture that intrigued me was given by a young Russian mathematician, Vladek Pucknachov, who proved the existence of a certain class of steady free-boundary fluid flows. To begin with, I had difficulty in understanding Pucknachov's result, but I realised it could be subjected to experimental verification. So on my return to Cambridge, I carried out an experiment that eventually led to my paper *"Behaviour of a viscous film on outer surface of a rotating cylinder"*, published in 1977 in the *Journal de Mécanique*. Here is the opening paragraph of this paper: *"It is a matter of common experience that if a knife is dipped in honey and then held horizontally, the honey will drain off; but that the honey may be retained on the knife by simply rotating it about its length. The question arises: what is the maximum load of honey that can be supported per unit length of knife for a given rotation rate?"* An answer to this question was given in the paper, but, more importantly, an instability leading to the phenomenon of 'syrup rings' was identified, which has led to many subsequent investigations.

Syrup rings on a rotating cylinder

Khartoum 1974

In 1974, Ibrahim Eltayeb, former PhD student at the University of Newcastle, invited me to act as External Examiner for the MSc in Fluid Dynamics at the University of Khartoum, where he was now a Lecturer. This was an exciting assignment and my first visit to Africa. The plane to Khartoum landed on what felt like a flat portion of desert, and the passengers walked across the sand to a tiny immigration building. I was accommodated at the Grand Hotel, still colonial in character, with huge fans suspended from the ceiling and gently turning to provide a pleasant current of cooling air. I duly carried out my duties as Examiner, which were not arduous. Eltayeb took me on a trip by bus to his parents' home in Wad Madani, some 100 km south of Khartoum; I recall sitting in the bus in a rainstorm waiting for it to set off on this trip. "When does the bus leave?" I asked Ibrahim. "When it is full" he replied, as indeed it was soon after this exchange.

Mohamed El Sawi, a former student of Derek Moore at Imperial College, also took me to visit his extended family in Omdurman, not far west from the confluence of the Blue Nile and the White Nile. The men of the family were seated in a circle on carpets on the ground of a large mud hut, and the women were standing in a wider circle round the perimeter, serving chilled tea and sweet cakes. The well-appointed interior somehow remained quite cool despite the very hot weather outside. It was a wonderfully hospitable welcome in a culture so distinct from any European lifestyle. At the market outside, I bought a Sudanese carving and an earthenware jug, good for the cooling of wine; also a turban and a flowing white Jellabiya, which astonished my neighbours in Barton Road when I wore it on my return to Cambridge.

Beersheva 1974

In 1974, I attended the first Beersheva Seminar on *Magnetohydrodynamics and Turbulence*, at the Ben Gurion University of the Negev, invited by Herman Branover, a refusenik from Latvia who had recently been allowed to emigrate to Israel from the Soviet Union, after several years of persecution and on payment of a 'ransom' to the Soviet authorities. This was my first visit to Israel. Beersheva, in the Negev desert, is where, according to the Book of Genesis, Abraham dug his well and planted a tamarisk tree. In the Bedouin marketplace, I purchased a carved model of Abraham as a gift for my father-in law, the Rev. Dr David Stiven (former minister of the Church of Scotland in the parish of Inveresk and then in Iona and the Ross of Mull, and a life-long scholar of Hebrew). Following the seminar, the small group of participants moved first to Jerusalem,

Beersheva 1974; Arthur Shercliff extreme right; I am third from right; René Moreau second from left

where we walked round the city wall, and noted that, as a city, it was indeed 'compactly built together'. We walked the Via Dolorosa, with its Stations of the Cross, and explored the four quarters of the city. In all this, I very much enjoyed the company of Arthur Shercliff.

We were then taken on a bus tour of several days, first to Jericho on the road taken by the Good Samaritan (a good parable for modern times), then to the legendary fortress of Masada on the coast of the Dead Sea, on which we later floated for relaxation. Then to Bethlehem and on up to the Sea of Galilee, familiar to me from the childhood experience of Sunday School tedium, but which now sprang to life in a more vividly evocative atmosphere.

Back to Trinity

In December 1974, the Senior Tutor of Trinity, Gareth Jones, was elected Downing Professor of the Laws of England. In those days, a University Professorship could not be held together with a College Tutorship, so the Senior Tutorship was suddenly and unexpectedly vacant. Having served for four years as a Tutor, I was in line to become Senior Tutor, a position that I accepted with some trepidation. Sure enough the position was far more demanding than I could have anticipated. For one thing, the issue of admission of women was coming to a head, and a College Meeting (of the Fellows) was held to vote on the necessary change of Statutes. It fell to me, as Senior Tutor, to propose the motion to change the Statutes, which was duly seconded and followed by an impassioned debate. In the course of this debate, James Lighthill, Lucasian Professor of Mathematics, argued so fiercely in favour of the change that he upset a number of die-hard Fellows who, adopting the manner of F.M.Cornford's 'principle of unripe time', considered that, while change might be good, no change was always better; in this, they could appeal to the Trinity motto *Semper Eadem* (always the same)! In the event however, the motion was passed with a commanding majority, enabling detailed planning to go ahead for the admission of women graduate students from 1976 and undergraduates from 1978.

College Council Minutes for 1975 reveal a further drama in which I, as Senior Tutor, was necessarily involved. This concerned two young mathematicians of the College, both *in statu pupillari*. Fifty years have elapsed since then, and the Freedom of Information Act allows me to tell this tale; but in order to protect the identity of these two, I shall simply denote them by the Greek letters, *Alpha* and *Delta*. It so happened that in those days, the undergraduate mathematicians of the College were wont to engage in a harmless pursuit, whereby each would attempt to break into the locked room of another while the occupant was out, overturn the bed and the bookcase and cause a degree of havoc, then leave and re-lock the door as if nothing untoward had happened. Well *Delta*, anticipating such an invasion, set up a booby-trap before going to dinner in Hall one evening. While he was out, *Alpha* successfully picked the lock of *Delta*'s room. As he pushed the door open, there was a minor explosion, which alerted the Porters, who, fearing a terrorist attack, called the Police. A Policeman duly arrived, entered the room, and switched on the light, triggering a second explosion, thus alerting the Cambridge Fire Brigade, who arrived forthwith and got the situation under control. The result was that *Delta* was charged with disturbing the peace by incendiary activity, while *Alpha* was not charged with anything, although it is arguable that he should have been. *Delta* was let off with a fine and a warning; these were tolerant days!

Paris 1975/6

But relief from the burden of the Senior Tutorship was at hand. I had previously arranged sabbatical leave in Paris for the academic year 1975/6, and that was sacrosanct. Denis Marrian, who was already experienced as a former Senior Tutor, took on the job in my absence. With the help of David Kelley, authority on Baudelaire and Trinity Fellow in Modern Languages, I rented a 5th floor apartment in Rue Molière just off the Avenue de l'Opéra. This apartment belonged to Desmond Ryan, Paris correspondent of an Irish Newspaper, who also owned a cottage in Moncourt-Fromonville, just south of Fontainebleau. Desmond and Mary were living in Moncourt, but would come up to Paris once a week or so, and look in at Rue Molière for lunch, so we remained in contact and became close friends.

I was based at the Université Pierre et Marie Curie ['Paris VI'] in Place Jussieu, which consisted of a set of massive towers in a hideous concrete setting. The Département de Mécanique was high up in Tour 66, accessible only by elevator, and this is where I sat, dealing with *JFM* correspondence which had followed me to Paris, and writing my monograph *"Magnetic Field Generation in Electrically Conducting Fluids"*, my project for the year. I also gave a course of lectures on the subject while there. By February 1976, I was up to Chapter 5 in a section on '*anti-dynamo theorems*', where I ran into a peculiarly intractable difficulty, which got me quite depressed. By good chance, I was invited by Hélène Lanchon to take a skiing break for a week at the end of February; this helped to blow all the cobwebs away. From then on it was plain sailing, and I finished the 12 chapters of the monograph just before returning to Cambridge at the end of July. The skiing holiday was refreshing in many ways, most notably through my friendship that developed with Jean-Pierre and Agnès Brancher, whose ineptitude on the ski-slope was almost as great as mine, so that we spent most of our time in the open-air cafeteria at the base of the slope, admiring the graceful descent of others in the party. Anyway, the whole week was a great joy, and totally revived my spirits. Thank you Hélène!

Around April that year, I received a letter from George Batchelor drawing my attention to a Chair at Bristol University, which was soon to become vacant on the retirement of Leslie Howarth (who had actually been Batchelor's PhD examiner, together with G.I.Taylor, back in 1948). I think George feared that the Senior Tutorship at Trinity would bring an end to my research career, and that a move to Bristol was the way to rescue me. In this, he was probably right! I applied for the Bristol job, was summoned from Paris for interview, and was duly offered the position, which I took up in January 1977.

Bristol 1977-1980

For the first six months in Bristol, I was in effect finding my feet in a strange Department of Mathematics, in which it soon became apparent that there were serious personality problems. I was placed in charge of the applied side of the department, and these problems were to loom ever larger during the three years that I remained in Bristol.

We bought a house on Upper Belgrave Road facing the Downs, and Linty and the children followed me to Bristol in July at the end of the school year. The boys then enrolled at Bristol Grammar School and the girls at Redland High School, where they were very happy. Our house was on four floors, with a basement that led onto a small paved garden with a beautiful vine that produced a good crop of grapes each year. I made wine from these grapes, and labelled the bottles (all four of them) *Château Haut Belgrave* and *mis en bouteille au château*, which seemed legitimate given our address. Only recently have I discovered that there is a genuine *Château Belgrave, Cru Classé Haut-Médoc*! So perhaps I committed a crime, but we drank all the wine at Christmas each year, so the incriminating evidence was quickly destroyed.

My colleagues in Bristol, Howell Peregrine, Philip Drazin and David Evans, were a wonderful support to me throughout my time there. And I had a great group of Research Students, and a very able post-doc, Brian Duffy, who moved on to the University of Strathclyde shortly after leaving Bristol.

Perugia 1978

Early in 1978, I gave a seminar at Heriot Watt University, and while there, Robin Knops asked me whether I could take over from him at a Summer School in Applied Mathematics, at which he had been teaching for several years. I readily agreed to this, particularly when he told me that this School was held at the University of Perugia, a beautiful city in the heart of Italy, and a wonderful location also for a family holiday! The school ran for four weeks through the summer, and we were able to drive to Perugia just in time for the start of the course. I lectured on fluid mechanics each morning, and we spent most afternoons on the Isola Polvese in Lake Trasimeno, about 20 km to the west of Perugia, where we found a delightful picnic spot, and where we all swam in the lake till sun-saturated and hungry for the open-air oven-baked pizzas for which Perugia was famous. We greatly enjoyed this glorious summer activity, which we as a family engaged in again the following year.

CHAPTER 5
The 80s: Political Tangles

The 1980s were the 'Thatcher years' in the UK, a period of belt-tightening all round, and particularly in the Universities across the country. This meant that when academic posts became vacant, they were not automatically refilled, but were instead usually 'frozen' for several years. Moreover many academics over the age of 60 were encouraged to take early retirement, to reduce the financial burden on their Universities. Of course this meant heavier teaching loads for those that remained in post. This was the situation when I returned to Cambridge from Bristol in 1980; it affected me particularly when Batchelor was induced to take early retirement in 1983 and I succeeded him as Head of Department. Funding was very tight, and Stephen Hawking's needs as Lucasian Professor were paramount.

But the political problems were far more widespread. The USSR, which I visited in 1982, 1985 and 1988 was slowly crumbling throughout the decade, culminating ultimately in the dissolution of the Soviet Union in 1991. And, by contrast, China, which I visited in 1986, was slowly emerging from its years of isolation under Chairman Mao (who had died in 1976), under the more liberal leadership of Deng Xiaoping. These developments were to have a profound effect on international collaboration in science. These 'political tangles' of the 1980s provided the background to this chapter.

My fortuitous return to Cambridge 1980

In 1980, an opportunity came to return to Cambridge, which I was swift to grasp. It happened in this way. I had invited George Batchelor to give a seminar at Bristol, following which we had him to a quick dinner at home. It had to be quick because George had to catch a train in time to get him back to Cambridge that evening. As I drove him in haste to Parkside station, I asked him about the vacant *Chair of Mathematical Physics (1978)* which was not yet filled. I knew that Ed Spiegel had been offered the Chair, but after much fruitless negotiation had declined it. George told me that the Electors were now at loggerheads, being unable to agree on either of the two leading candidates. In these circumstances, I asked George if there was any point in my applying, and he immediately encouraged me to do so. He caught his train at Parkside with moments to spare.

That evening, I wrote out my cv by hand and faxed my application to George the following morning. Within a matter of weeks, he had managed to gain enough support to convince the Electors that I was a viable choice, and they seized the opportunity to reach agreement on me, as an obvious compromise candidate.

And so, at the age of 45, exactly half the age I am now, I returned to Cambridge in the summer of 1980 as the first holder of the above Chair, and I was re-elected to a Fellowship at Trinity College. For my first year back in Trinity, I occupied a first-floor room on E staircase, Great Court, that Isaac Newton had used as his study; it had been vacated that year by its normal occupant, the economist Stephen Satchell, who was on leave. The room has a corner door to a spiral staircase that leads to the roof of the Great Gate, where Newton mounted his telescope to observe the motion of the planets.

By 1980, Professors were permitted to do a limited amount of College teaching, and I took advantage of this freedom; but students were so over-awed by Newton's room, that they were quite unable to concentrate, so perhaps it wasn't such a good idea to be there after all.

On a visit to Tokyo for a conference some years later, Linty and I found ourselves in a packed underground train carriage. A local inhabitant was eager to engage in conversation with us, so Linty told him that I had worked in Cambridge in the room that had been occupied by Newton. He was so impressed that he shouted this information in Japanese to the whole carriage, whereupon all the occupants bowed simultaneously in our direction and applauded enthusiastically. The name of Isaac Newton certainly resonates, as we see from the crowds of tourists who throng daily around the apple tree on the little lawn to the right of the Great Gate.

Hester's lymphoma crisis 1982

Early in 1982, our elder daughter Hester, by now aged 15, had a very troublesome cough and was seriously ill; she was eventually diagnosed with Lymphoma. She had to undergo an extensive course of chemo- and radiotherapy in the two years that she was studying for A-levels. We gave her a moped so that she could get easily to Addenbrooke's hospital, and retain her independence throughout this dreadful process, which she came through with flying colours. She achieved entrance to Exeter College Oxford in 1985 to read Modern Languages (French and Modern Greek). Coincidentally, she found that her Tutor was the same Jim Hiddleston, who had been my classmate at George Watson's College for 10 years, 1943-1953. In her second year at Oxford, Hester met her husband-to-be, Fred Tingey; they married in 1991, and have four children (now all adult), Chloe, Tabitha, Alfie and Bathsheba.

USSR 1982

Soon after my return to Cambridge, I received a letter from G.I. [Grisha] Barenblatt in Moscow, inviting me to spend a month in the USSR, under a USSR-Royal Society exchange agreement. I immediately accepted this invitation, and spent the next year endeavouring to learn a little Russian. This was possible through courses offered by the Cambridge City Council. I recall applying to attend one of these courses. I presented myself at the appointed time and place, after a day teaching mathematics. It was raining and I was wearing an old Mackintosh that Linty had found for me at the Oxfam shop, so I looked quite disreputable. I was asked what O-levels I had, and I had to admit none (having taken only the Scottish Leaving Certificate). The officials looked askance at each other as if to say, what have we here, and then asked what was my job. I said I was a Professor of Mathematics, at which they evidently thought that I was quite mad, but they let me enrol anyway, as the class was evidently going to be a very small one. So, yes, I took the course and actually ended up with a B at O-level, which I felt was only barely deserved. But this elementary level of Russian was to stand me in good stead when I set off for Moscow in March 1982 for a four-week visit.

Grisha had arranged everything with meticulous care and attention to detail. The country was still under the firm control of Leonid Brezhnev, with as yet no hint of *Glasnost*. It was still a regime surviving under military and police control that was everywhere in evidence. The Soviet economy was at that time going from bad to worse, and one could sense the widespread disillusion of the people with the oppressive system that controlled their lives.

My journey took me to Novosibirsk, Irkutsk, St Petersburg (or Leningrad, as it then was) and Riga, as well as Moscow. I visited 12 different research Institutes and gave 12 lectures. The questions raised during these lectures were challenging, and I soon learnt to appreciate the Soviet tradition, whereby a 'seminar' based on a one-hour lecture could well extend to two or even three hours through animated audience participation.

My host in Novosibirsk was Vladislav (Vladek) Pukhnachov, whom I had first met in Trieste in 1974. He met me at the airport in a blackout due to power failure, and took me in a battered Siberian Academy car to Akademgorodok, the site of no less than 35 Research Institutes of the Soviet Academy of Sciences. I stayed for a week, based at the Lavrentyev Institute of Hydrodynamics, where I met Vladimir (Volodya) Vladimirov, with whom I have enjoyed close friendship ever since. The temperature in Akademgorodok was -20 C throughout my stay, with hard-packed snow on the ground and crisp sparkling weather. My hosts (Vladek and Tanya Pucknachov) taught me the elements of cross-country skiing at the weekend and took me to the Novosibirsk Ballet one evening. They entertained me at their home with caviar and Armenian cognac, Siberian ravioli and rowan compôte; all most delicious!

Vladek and Tanya Pucknachov in 2011

On Saturday 20 March, Vladek accompanied me by plane to Irkutsk, where we were met by Sam Vainshtein (who, with Zel'dovich, had introduced the 'fast dynamo' concept). The next day, we drove to Lake Baikal, still frozen with two metres of very clear ice, and witnessed the arrival of a long thin line of skiers completing a 40 km annual challenge ski-run across the lake.

Vainshtein worked at SibIZMIR (the Siberian Institute of Geomagnetism, Ionosphere and Radio Wave Propagation). I lectured there with some difficulty, because there was no overhead transparency projector, the blackboard was primitive, and the chalk was a lump of calcite newly hewn from local deposits. The discussion was however animated and altogether it was an exhilarating encounter.

Back in Moscow, I had stimulating discussions over the next few days with Arnol'd and Zel'dovich. The timing was perfect because it gave me advance knowledge of the book *Magnetic Fields in Astrophysics* by Zel'dovich, Ruzmaikin & Sokoloff, published one year later in 1983. It was here that Arnol'd passed me his 1974 paper [in Russian] that he wanted me to read; I had it translated when I returned to Cambridge, and read it, but didn't recognise its significance till three years later.

Stanislav Braginski in Cambridge en route to USA, 1988

I also visited the geophysics institute where Stanislav Braginski worked, and had a good discussion with him, marred by the fact that a severe-looking lady, obviously KGB, sat in the corner of the room throughout our meeting, clearly to ensure that Stanislav showed me only his already published work, and could not reveal any current activity. Braginski emigrated from the Soviet Union in 1988, as soon as the political climate allowed, and settled finally at UCLA through the good offices of Paul Roberts, who had himself emigrated to the USA, on his retirement at Newcastle in 1986.

On Friday 26th, following a memorable dinner at his home, Zel'dovich put me on the night train to Leningrad, with a sly parting wink, for I found myself sharing a sleeping compartment with a beautiful *devochka*. It occurred to me that this might be a honeypot, an uncharitable thought because I arrived in Leningrad unscathed the following morning! There I was met by E.M. Drobyshevski who took me to the huge modern Hotel Moskva overlooking the River Neva on which great blocks of ice were still floating down towards the sea.

It was a particular pleasure to renew acquaintance with Olga Ladyzhenskaya, who was my guide to Pushkin and Pavlosk, and at the Ermitage over the weekend. I lectured at the Mathematical Institute two days later on the question of whether the Euler equations do or do not develop a singularity within a finite time. This is a question that remains open to this day; massive computational power has been thrown at the problem, but has still failed to provide a definitive answer, yes or no. Discussions with Ladyzhenskaya and V.A.Solonnikov were thought-provoking, as they had an original approach to existence problems of this kind.

I went on by train to Riga, capital of Latvia, still then of course part of the Soviet Union, although the Latvian spirit of independence was very evident in the intensely nostalgic Latvian songs sung by choirs in local costume. Here I visited the Riga MHD laboratory in Salaspils, at that time the best of its kind in the world. My host was Olgerts Lielausis, and I was interested to meet also Agris Gailitis, leading dynamo theoretician, who was pioneering the experiments in a closed circuit of liquid sodium that eventually led to the 'Riga dynamo'.

My final week was spent back in Moscow at Barenblatt's Institute of Oceanology, and at the Institute of Atmospheric Physics, where A.M.Obukhov, famous student and colleague of Kolmogorov, showed me his experiments in which vortices are excited in a layer of electrolyte by 'Lorentz manipulation'. Here I met Akiva Yaglom and Georgi Golitsyn; also Evgenyi Novikov, who was to defect to the USA in dramatic circumstances the following year during an IUTAM Symposium in Kyoto.

Grisha Barenblatt was wonderfully supportive at every stage, taking me at the weekend to his dacha in Abramtsevo and to the ancient monastery at Zagorsk, and later to the heavily guarded Novodyevichy cemetery in Moscow, where politicians, Academicians, composers, and generally the Soviet elite, are laid to rest, each tombstone bearing an appropriately chiselled bust. [Arnol'd is now interred in this cemetery.]

I returned to Cambridge after this month of non-stop scientific, cultural and social activity, exhausted and exhilarated in equal measure.

Head of Department 1983

It was almost as cold in Cambridge in January 1983 as it had been in Novosibirsk. The Cam froze making it possible for me to skate to work. This was the last year when this healthy means of propulsion has been possible in Cambridge, a sure sign of climate warming.

And it was then that George Batchelor dropped a bombshell: he announced that he was going to take early retirement, having served as Head of Department for 24 years. He was a victim of

Skating to work January 1983

the severe cuts in funding of the Thatcher years. He was, he said, 'made an offer that he couldn't refuse'. His Professorship was to be frozen for five years; this was how the University was responding to political pressures.

The Department had to choose a new Head, and soundings were duly made. There was ultimately a vote and I was the lamb to the slaughter. I was to serve as Head of Department for eight years, but for one of these (1986/87) I was on sabbatical leave, and John Taylor, who had succeeded John Polkinghorne, nobly stood in for me as Deputy Head for that year.

The first thing I did was to have a bubble tube installed in the Department. It ran from the basement up through the Common Room, then on up through the library on the first floor to the second floor landing. The tube was filled with silicon oil, and 'wavy skirted bubbles' released in the basement travelled up through the four floors of the building; very satisfying from a fluid dynamical point of view!

Penelope and Pierre Comte, 1989 in the DAMTP Common Room by the bubble tube. A wavy skirted bubble rises through the tube.

Stephen Hawking had succeeded James Lighthill as Lucasian Professor in 1978, and was well established in a ground floor office of the Department, just off the Common Room. His speech had become very slurred by this time, and communication was difficult. I did what I could to help him in applying regularly to Leverhulme for continuation of the support that he desperately needed. There was a crisis situation in 1985, when Stephen contracted pneumonia while at a meeting in Switzerland and was flown back to Addenbrooke's Hospital by air ambulance. After a tracheotomy, he could no longer speak, and the situation appeared desperate. I visited him, and found it hard to believe that he could recover. But Stephen's indomitable will-power, and the vital support of his wife Jane, got him through the crisis. When I visited Stephen in hospital, they were already in discussion with computer experts who were keen to provide the voice synthesis hardware that would provide Stephen with the voice that would be recognised world-wide for years to come. The necessary computers were subsequently provided by Gordon Moore of Intel. Stephen's *Brief History of Time* was published two years later, and the financial demands of round-the-clock care were soon more easily met.

Fergus: a doubly manic episode

I was giving a Part III course of lectures in Cambridge during the Lent Term 1983 when Fergus, now aged 21, had one of his manic episodes. It was about 6 a.m. on a Monday morning that we received a phone-call from a friend in Loughborough whom I had met at a Conference some years previously. He said that Fergus had appeared at his doorstep in a highly excited state the previous evening, had stayed but had not slept all night; and could I come to collect him and take him home to Cambridge. I was due to lecture at 10 a.m. but of course set off immediately for Loughborough, about 90 miles from Cambridge, thinking I could get back in time to give my lecture. This was not to be, for when I found a parking place in Loughborough beside a churchyard near my friend's home, I was distracted by a querulous cry "help me, help me" from what turned out to be an old lady wandering among the tombstones. It was still dark and a cold damp morning. What could I do but respond? I took her into my car -- she was in night attire and shivering with cold -- but she was incoherent and didn't know where her home was. A milk-float delivery man came to the rescue, he knew the lady well, and gave me her address, so that I was able to deliver her safely back to her elderly husband from whom she had wandered -- apparently a not infrequent occurrence. I lost so much time over this that I had to phone Cambridge to cancel my lecture that day. My class was very understanding when I explained the situation at the following lecture two days later. In the meantime, I collected Fergus, who, vaguely aware of his own manic state, had abandoned the car (our old VW Dormobile) that he had been driving, he couldn't remember quite where, and was in grave need of calming medication. A very anxious day followed as we managed to sort things out and finally get back home to Cambridge.

Penelope in Nice and Susa, 1983

In August 1983, Linty and I spent two weeks at the Observatoire de Nice, invited by Uriel Frisch to engage in discussions on turbulence. We travelled by car with our younger daughter Penelope, then 15 years old, via Susa in Italy, where Penelope made friends with Paulo, a boy of a similar age. Paulo's mother Lilliana invited Penelope to stay on in Susa for a few days, while Linty and I continued the journey to Nice. Some days later, we received a message that Penelope and Paulo were hitch-hiking to Nice. They didn't arrive when expected, and Linty and I became increasingly anxious. Eventually, I decided to drive back to Susa, a four-hour journey, to see if they were still *en route*, communication being otherwise impossible. I found no trace of the runaways, but was welcomed by the charming Señora Lilliana, who insisted I stay the night. I returned to Nice the next morning, by which time Penelope and Paulo had found their way to the Observatoire, having slept overnight on the beach in Nice by the Promenade des Anglais. Oh, the parental anxieties induced by teenage daughters!

Palermo, January 1985

One of my students in Perugia, with the wonderful name Pantaleo Carbonaro, invited me some years later to spend a month at the Universita degli Studi di Palermo. Who could resist such an opportunity? I accepted the invitation for the month of January 1985, and went with Linty and Penelope who was by then about to study French and Italian at Oxford. We rented a cottage at Mondello, a few miles to the west of Palermo, and conveniently near the bus terminus. Fortunately it was January rather than July, and the blustery weather suited us well. One could however sense a rather unsettling Mafia presence in the market place!

Salvio Mercurio very kindly took us one day to Erice and to the Greek temple at Segesta, completely deserted at that time of year. Salvio was rather in favour of the Mafia, which provided protection, at a price, for local businesses. He may have changed his mind when world headlines reported the 1992 assassination by the Mafia of the Palermo Judge, Giovanni Falcone, by means of a car bomb concealed on a highway not far from Mondello!

The peace and quiet of Mondello gave me the ideal opportunity for research in the afternoons, fortified by lunch by the sea over a good bottle of Sicilian white wine. It was here that I began to understand the process of magnetic relaxation, in effect the counterpart of dynamo theory, which allowed me to argue that magnetostatic equilibria of arbitrary field topology must exist in a perfectly conducting fluid; and then, arguing by analogy, that similar steady 'Euler' flows of arbitrary *streamline* topology must exist in an ideal fluid. In great excitement, I wrote a paper on this subject. It was only then that I recognised the significance of the paper that Arnol'd had given me in Moscow three years earlier. I referred to this paper in the opening sentence of my paper that I submitted to *JFM* just one week after our return to Cambridge: "*In a paper of great fundamental interest . . . , Arnol'd (1974) has posed the following problem in magnetohydrodynamics . . .* ", and with a footnote "*Arnol'd attributes the conception of this problem to Ya. B. Zel'dovich*". I was elected to the Royal Society in 1986, largely, I believe, on the basis of this paper.

Lavrentiev Readings, Kiev 1985

In September 1985, I attended a meeting in Kiev, one of a series of '*Lavrentiev Readings*' held annually in the Soviet Union. This gave me an opportunity to present my paper on topologically complex flows to a Russian and Ukranian audience. I was of course eager to meet Vladimir (Dima) Arnol'd again, and I arranged a few extra days in Moscow with this in mind. It was not going to be easy, because Arnol'd lived outside the city limits, and contact with him by 'foreigners' was discouraged, to say the least. Moreover, a KGB 'minder', in the person of Tania Pucknachov, was assigned to accompany me to Moscow and to be my guide, in other words to report to higher authority on my movements and contacts throughout my visit. However, on one day, I was able to give her the slip. I got up very early, bypassed the 'dejournaya' on the hotel landing, and took the Moscow Metro to its terminus, and then a bus as far as I could towards Dima's address. But then I was stuck, as I had no detailed map of that neighbourhood. Here, my rudimentary Russian was indispensable. A milk float had stopped at the bus stop, so I showed the milkman the address, and I understood his Russian invitation to '*Jump in, I have to deliver milk there and I can take you*'. And thus it was that I arrived at Dima's apartment, where we had an animated discussion. At my suggestion, Dima published an English version of his 1974 paper the following year in the journal *Selecta Mathematica Sovietica*.

I invited Arnol'd to give the Rouse Ball Lecture in Cambridge (a lecture intended to be accessible to a mainly undergraduate audience). We arranged this for November 1988, when it became possible for him to get a visa to travel to the UK. He arrived in Cambridge on a Saturday, the lecture being scheduled for the following Monday morning. I invited him home on the Sunday and took him a walk by the tow path down the River Cam for lunch at Clayhithe. When we got there, he decided to have a swim before lunch, stripped naked and plunged into the river. He swam across and back, shook himself dry, dressed and declared himself ready for lunch. It fell to me to introduce him to a large audience the following morning. I naturally summarised his brilliant record of research, but when I added that he had swum across the Cam the previous day, a cold November morning, it was this that got the greatest round of applause.

V.I. Arnol'd, before his swim in the Cam, November 1988

Feynman and the first Dirac Lecture

Richard Feynman gave the first Dirac Memorial Lecture on 16 June 1986. He had flown from California, and was at St John's College for the duration of his short stay in Cambridge. Feynman was a member of the Rogers Commission that had exactly one week earlier published its report on the Space Shuttle Challenger disaster, which had occurred in January that year. I collected Feynman from St John's College and walked him along via Free School Lane to the Babbage Lecture Theatre, where the lecture was to be delivered. He was still jet-lagged, and I was concerned that he would have difficulty in getting onto the stage, far less delivering a one-hour lecture. In the event, he gave a brilliant lecture *"Elementary Particles and the Laws of Physics"* before a packed lecture theatre. [This lecture was video-recorded and can be found on YouTube].

That evening, the theoretical physicists of Trinity College entertained Feynman to dinner. Over the Port that followed, he regaled the gathering with his experience on the Rogers Commission in demonstrating how the O-rings had been the root cause of the disaster. He was in great good-humoured form. Tragically, he was to die of cancer less than two years after this event, in February 1988.

China, September 1986

I spent three weeks of September 1986 in China at the invitation of the Chinese Academy of Science (Academica Sinica). The economic reforms initiated by Deng Xiaoping were by then well underway, and China was beginning to open up to academic visitors from the West. My first meeting with scientists from China had been earlier that year, at the meeting of the General Assembly of IUTAM, held in University College London at the invitation of Sir James Lighthill, then President of IUTAM. China was represented at that meeting by Zheng Zhemin, Director of the Academy's Institute of Mechanics, and Wang Ren, Chairman of the Department of Mechanics at Peking University; their presence at this meeting had made a dramatic impression.

The 1980s had been a period of extraordinary developments in China, and I was lucky to be invited at such a critical time of change. I was accompanied on this journey by my mother Emmeline, then aged 78, who had always longed to visit China, and who jumped at this opportunity. She kept a detailed journal of the visit, and I have been able to draw on this in the following account. The Chinese have a great respect for 'senior citizens', the more senior the better, and this was very evident in the great courtesy with which we were welcomed and treated throughout our travels. As regards the various sight-seeing trips, I will give here only the briefest summary.

Our visit started and ended in Beijing, and included a round trip to Hangzhou, Shanghai, Chengdu, and Xian, where I visited Universities and gave lectures. We were welcomed in Beijing by Professor Wenrui Hu of the Institute of Mechanics, and escorted to the Friendship Hotel, which was to be our base. The following day being a Sunday, we were escorted by Wenrui's colleague Jai-Fu to the Temple of Heaven, the first of several sight-seeing trips that had been pre-arranged. On the way back, we paused at Tiananmen Square, a vast area of great significance in Chinese history.

With my mother, Emmeline, on the Great Wall, Sept. 1986

On the Monday and Tuesday mornings, I lectured at the Institute of Mechanics, where I was happy to renew acquaintance with Zheng Zhemin. In the afternoons we were escorted to the Forbidden City and the Summer Palace; and on Wednesday on that tour of tours to the Great Wall on which 300,000 men had toiled for 10 years, and the Ding Ling Tomb (one of the Ming Tombs) in the bowels of the earth, described by Emmeline as *"a subterranean stone-built palace, a grim and fearful place indeed with its side alleys and now empty sarcophagi. The one cheerful thing about the Ding Ling Tomb is its name!"*

On the Thursday, we flew to Hangzhou, where we were accommodated in the University Guesthouse, and where we were guided in succession to *"the most beautiful lake, the largest Buddhist temple, the highest Pagoda, the first Communist iron-built road bridge, the finest tea made with the purest (Dragon Spring) water"*. It rained incessantly throughout our three days here, leaving us with a somewhat soggy impression of superlatives before we continued by train to Shanghai, where we were met by Professor Dai, of the Shanghai University of Technology (as it then was).

On the following day, I was taken on a tour of the University, first to the Computer Science Department, where an IBM 4167 had been recently installed; then to the Library where, according to Prof. Dai *"we take all journals"* -- since 1984, that is; third, to the Department of Mechanical Engineering, where the speciality was high precision optics applied particularly to the problem of detecting stress patterns in elastic bodies subjected to bending and twist; and finally to the Institute of Applied Mathematics and Mechanics (where I lectured), actually a small house formerly owned by a priest, now containing a 10-month old VAX computer with 15 terminals. Here I met Professor Cai, distinguished by the Mao jacket and sandals that, as I was told, he always wore. This Institute had plans to move in 1987 to a new five-storey building with a fluid mechanics lab in the basement.

We then flew to Chengdu, the capital of Sichuan Province in South-West China, and the gateway to Tibet. Here I was welcomed by Professor Han of the Institute of Mathematics and Physics, one of 7 or 8 Institutes of the Chengdu Branch of the Chinese Academy of Science. His small group of five graduate students and four assistants focused mainly on non-Newtonian fluid mechanics, with interest also in drag reduction and wave surges in lakes caused by landslides. I gave two lectures here on slow viscous flow, which seemed the appropriate topic. We were accommodated at the new guesthouse of the Economics Institute, simple but adequate, and enjoyed excellent spicy Sichuan-style food.

Then on to Xian where, under the guidance of Li Ou of the Shaanxi Astronomical Observatory, I was taken to see the famed terracotta warriors, discovered in 1974 and excavated since then. It has to be seen to be believed! An amazing relic of an ancient civilisation.

And finally we returned to Beijing, where on 24th September, I visited the Department of Mechanics of Peking University, at the invitation of its Chairman, Wang Ren. Here, they are well equipped with wind tunnels, water channel, shock tubes, etc. The Department admits 70 students each year, of whom 30 ultimately stay on for graduate work.

I gave my final lecture at the Institute of Mechanics, and had interesting discussions on astrophysical fluid dynamics with WenRui Hu, who had been much influenced by C.C.Lin when he returned to China in 1975 to lecture on his controversial work on density waves in explaining the structure of spiral galaxies. [I also had been much influenced by C.C.Lin, through his 1955 monograph on *Hydrodynamic Stability*, which I had studied in preparing my Yeats prize essay at Trinity College in 1959.]

It was a great honour and privilege to be entertained to dinner that evening in one of the banqueting rooms of the Great Hall of the People in Tiananmen Square by Academician Zhou PeiYuan, President of Peking University, whose early work on turbulence was known to my own mentor George Batchelor. This was a fitting climax to these three wonderfully memorable weeks.

MADYLAM (INPG Grenoble) 1986/7

René Moreau, a leading authority in liquid metal magnetohydrodynamics, arranged a 6-month visit for me to MADYLAM (the Laboratoire de Magnétodynamique of the Institut Nationale

Polytechnique de Grenoble (INPG)) that he had founded some years earlier. During my stay at INPG, I gave a graduate course on turbulence and dynamo theory; one of the students in my class was Pierre Comte, who was later to become partnered with my younger daughter Penelope. I shared an office with Sherwin Maslowe, who was visiting from McGill University; I met him again 25 years later at an IUTAM Symposium in Fukuoka, Japan, where we shared happy memories.

We rented a chalet at St. Hilaire du Touvet on the Plateau des Petites Roches above Grenoble, from René's technical assistant, Robert Bolcato, who became a close friend through that cold winter. He taught us in October how to make cider from a lorryload of fallen apples; it was ready to drink by Christmas when the family came to visit. Our most lofty achievement that winter was to climb the snow-clad Dent de Crolles.

The University of California at San Diego (UCSD) 1987

I was attached for three months (April-June 1987) to The Institute for Geophysics and Planetary Physics (IGPP) on a Green Scholarship, and for a further two months at the Institute for Nonlinear Science (INLS), which had just recently been established under the Directorship of Henry Abarbanel (1943--2023). IGPP is wonderfully placed by the La Jolla beach, and we took full advantage. My host was Hassan Aref,

Hassan Aref (on the left) chats with me and Grisha Barenblatt, at ICTAM Chicago, 2000

with whom I had a close working relationship. As Secretary of the Congress Committee of IUTAM, I was busy in 1987 preparing the programme for the 17th ICTAM to be held in Grenoble in August 1988. (Hassan was later to take on the same role of Secretary of the Congress Committee from 2004 until his untimely death in September 2011.)

During this period, I was invited by John Kessler to give a seminar at the University of Arizona in Tucson. I spent a few days there where it was very hot, c.40 C, but a dry heat, so tolerable. John took me for a walk in the desert to view the enormous cacti.

At INLS, I met Michael Freedman who had the previous year been awarded the Fields Medal for his work on the Poincaré conjecture. He took an interest in my presentation of the magnetic relaxation problem, and later proved that the magnetic energy of any topologically nontrivial field has a positive lower bound; this had been known for fields of non-zero helicity, but there are many examples of 'topologically nontrivial fields' of zero helicity, for example fields with the Borromean-ring topology (three rings which are triply linked although no two of them are linked). For such fields, the 'positive lower bound' result may be physically obvious, but to prove its existence required the mathematical skill that Freedman was able to bring to bear on the problem.

John Kessler in the Tucson desert

Fergus, deceased 1987

We returned to Cambridge in September 1987. Our son Fergus had suffered a lung collapse and had been unable to visit us in California. He came home from Windsor, where he had been working, in a deep depression, and tragically committed suicide on 16th September, two days before his 26th birthday. The lament of Chidiock Tichborne conveys better than I can myself something of the sorrow of the situation:

Fergus, in his prime

My prime of youth is but a frost of cares,
  My feast of joy is but a dish of pain,
My crop of corn is but a field of tares,
  And all my goods are but vain hope of gain.
The day is past, and yet I saw no sun;
  And now I live, and now my life is done!

....
My glass is full, and yet my glass is run;
  And now I live, and now my life is done!

Patras 1988

In June, Linty and I flew to Thessalonika, in part to see our daughter Hester, who was completing her year at the University there, and partly to attend a fluid dynamics Symposium in Patras. We rented a car and drove down the Aegean coast, then inland via Larissa and on to Patras, taking advantage all the way of Hester's fluency in the Greek language. On the way back, we stopped at Delphi and marvelled at this ancient 'navel of the world'.

Hester and Freddie visited Greece frequently, and much later bought a seaside cottage on the peninsula of Lefkada, thought by some to be the Ithaca of antiquity, where we enjoyed delightful holidays in 2008 and 2009.

Riga 1988

In 1988, I again visited the Soviet Union, in order to attend a meeting in Riga on Magnetohydrodynamics. For this it was necessary to fly with stopover in Moscow, where I was met by Agris Gailitis, who had flown from Riga to meet me. It emerged that he had an ulterior motive, for in the few hours before our flight to Riga he took me to a large supermarket to buy sugar. Why, I asked, did he want to buy sugar in Moscow? He replied that it wasn't available in Riga, but he had harvested plenty wild fruit there, and he needed sugar to make jam. So we found our way to the supermarket, a hall the size of a football pitch with completely empty shelves; but in the far corner a queue had formed, and we joined this. Everyone in the queue was buying sugar, nothing else being available for sale. Agris bought 2 kg (the maximum allowed), and I bought the same on his instruction. We left the hall and stored the sugar in his large rucksack. Then to my surprise, he insisted that we join the queue again and repeat our purchases. So finally we returned to the airport with 8 kg of sugar, a heavy load for Agris, but he was very pleased to be burdened in this way!

The meeting in Riga was quite charged, partly with a sense of optimism that control from Moscow was weakening, and that independence might soon be realised, and partly with a sense of foreboding that there might still be a severe clampdown at any moment. Extreme caution on all fronts was still obligatory. But the meeting was a good one in which ideas flowed in both directions. One evening we were entertained to choral singing of a fervency that was extremely moving.

Austria, January 1989

Early in 1989, Linty and I spent a week in Austria, where I had been invited to give lectures at the Universities of Vienna, Graz and Innsbruck. In Vienna, we stayed at guest accommodation in the grounds of the old Observatory. It was in Vienna that we first discovered that delicious combination of tomato and mozzarella with a drizzle of herbs and fine olive oil, a favourite ever since! In Innsbruck, we were taken by cable car up to a ski resort, where we were caught in a white-out -- impossible to see more than one metre in any direction -- quite scary, but our guide kept us on the path of righteousness!

The Abutilon miracle

Soon after our return to England, I visited Eric James in London. We talked about our loss of Fergus, and I told him of the miracle that he, Eric, related in a subsequent '*Thought for the day*' on Radio 4. Fergus was a gifted gardener, and had given me a potted plant -- an Abutilon -- two years before his death, which I kept in my office at DAMTP, and assiduously tended, but it had never flowered. But the day after Fergus's suicide, it burst into bloom --- a host of beautiful bell-shaped flowers; it was as if Fergus was telling me through this Abutilon that he would live on in another dimension, as he has done in our fondest memories over the years.

IUTAM Symposium: Topological Fluid Dynamics, Cambridge 1989

'Topological Fluid Dynamics' had taken birth during the 1980s, and the time was ripe for the IUTAM Symposium that I organised in Cambridge in 1989, in collaboration with Arkady Tsinober [1937--2020], a refusenik from Latvia who had managed to move to Israel in 1978. Our Scientific Committee included V.I Arnol'd (USSR), U. Frisch (France), H. Hasimoto (Japan), F. Hussain (USA) and E.N. Parker (also USA), and a galaxy of stars were included in the programme of lectures. The Proceedings were published by CUP in 1990. On the social side, with the help of Konrad Bajer, who was that year completing his PhD under my supervision, we organised a 'punt picnic', this being a very practical fluid dynamical experience for all participants; and fortunately, the weather held good for the occasion.

Participants at the IUTAM Symposium "Topological Fluid Dynamics", Cambridge, 1989

Disembarking for the punt picnic

Olga Ladyzhenskaya and Sasha Ruzmaikin (on the right) with others at the punt picnic

CHAPTER 6

The 90s: The Ravelled Sleeve of Care

Tight knots

Stephen Hawking permitted himself just one equation (which had to be $E = mc^2$) in his blockbuster *Brief History of Time,* on the grounds that any additional equation would halve the number of sales of the book. I shall be similarly restrained. But even so, a little (very little) mathematical background is required.

Consider a closed curve C in three-dimensional space, which may be knotted, but which does not intersect itself. Now imagine that a tube (like a hosepipe) surrounds this curve. In magnetohydrodynamics, it is natural to think of such tubes as 'magnetic flux tubes' imbedded in an electrically conducting fluid. This imagery would have been quite natural to the great Scottish physicist and mathematician, James Clerk Maxwell [1831--1879], whose theory of electromagnetism was as revolutionary in its time as it is relevant today for all devices dependent on wi-fi technology.

Maxwell understood that a tension exists in magnetic flux tubes, just as if they were made of stretched elastic material. Thus, if released from rest in an ambient perfectly conducting fluid, a knotted tube will always tend to contract in length, and this contraction will continue until the tube comes into contact with itself -- a 'topological constraint' related to the knot structure. Maxwell may have also recognised that such a flux tube is endowed with a 'magnetic energy' that must decrease as the contraction proceeds. This energy must reach a minimum when the contraction finally ceases; at this stage, the knot is as fully tightened as it can be.

So we may ask the question: for any given knot K (which may be the trefoil, or the figure-of-eight knot, or the bowline, or any of the thousands of knots that have been classified --- a wonderful playground for mathematicians!), for any given knot, what exactly is this minimum energy when it is in this tightened state?

Well, during the contraction process, three things remain constant if the ambient fluid is incompressible and perfectly conducting: the magnetic flux Φ in the tube, its volume V, and the helicity $H = h\,\Phi^2$, which depends essentially on the twist of the magnetic field lines within the tube (which behave like the internal strands of a knotted rope). The minimum energy M_K for the knot K can depend only on these conserved quantities, so on purely dimensional grounds must take the form

$$M_K = m_K(h)\,\Phi^2\,V^{-1/3},$$

where $m_K(h)$ is simply a number (for each choice of h) that can be computed. This is the one equation that I permit myself. A simple version of it appeared first in my 1990 *Nature* paper "*The energy spectrum of knots and links*". My student Atta Chui, subsequently computed the function $m_K(h)$ for various torus knots.

A closely related question is this: suppose we tie a particular knot K in a rope of length L of fixed cross-section A. Then what is the minimum length L_{min} for which such a knot can be tied, the two loose ends then being smoothly joined together (don't ask how) so that a tight closed knot tube is again formed? Much fascinating work has been devoted to this problem in recent years by Jason Cantarella (University of Georgia) and others.

Why then "the ravelled sleeve of care"? If you knit a sleeve of a jersey, it is actually unknotted: for if you remove the knitting needles and carefully pull the two loose ends of wool, it will completely unravel, so knitting is not knotting; but Shakespeare had poetic, not mathematical, license! And besides, "the ravelled sleeve of care" describes well the various responsibilities that I undertook during the 1990s.

The Isaac Newton Institute for Mathematical Sciences

Welcoming Sir Michael Atiyah as Master of Trinity College, 1990

Planning for the establishment of a visitor research institute in the mathematical sciences, similar to the Kavli Institute for Theoretical Physics (KITP) in Santa Barbara, had begun in 1989, and this planning intensified when Sir Michael Atiyah moved to Cambridge to succeed Sir Andrew Huxley as Master of Trinity College. The installation of Sir Michael as Master followed a centuries-old tradition. He was required to knock on the Great Gate and present his Letters Patent from the Queen to the Head Porter, who then closed the Gate and took the Letters to the Chapel for scrutiny by the Vice Master. These Letters being found to be correct, the Vice Master led a procession of Fellows to welcome Sir Michael, the Great Gate now being opened wide. There was then a great doffing of mortarboards and velvet caps as Sir Michael, followed by the Fellows, proceeded back to the Chapel for the formal admission ceremony.

Sir Michael also accepted appointment as Director of the proposed mathematical institute, and from this point on, at Michael's suggestion, it was named the Isaac Newton Institute for Mathematical Sciences (INI). A building was already on the drawing board and was to be built on land provided on an exceptionally long lease by St John's College. The Institute would host visitor research programmes, scheduled to start in July 1992. I was soon to find myself organising one of these programmes on the subject of Dynamo Theory; this would be held in parallel very appropriately with a programme on Knot Theory. But that was still some way into the future.

Batchelor's 70th birthday celebration

But I had other preoccupations at this time. George Batchelor's 70th birthday fell on 8th March 1990, and we arranged a two-day celebration in DAMTP to thank George (or GKB as he was known to everyone) for his lifetime promotion of Fluid Mechanics as a distinctive and vital branch of the physical sciences. We arranged a programme of lectures by a distinguished group of friends and colleagues, and we commissioned 34 papers that were published in the special Volume 212 dedicated to GKB, for which I acted as Guest Editor.

Batchelor's 70th birthday celebration at DAMTP; from left: Brooke Benjamin, Akiva Yaglom, Owen Phillips, Andy Acrivos, Grisha Barenblatt. GKB, HKM, Milton Van Dyke, Philip Saffman

My memory of this meeting is tinged with sadness, as it was soon after this that George was diagnosed with Parkinson's disease. Despite the very serious effects of this, both physical and mental, from which he suffered, he continued to walk daily to the Department until a year before his death just after his 80th birthday in the Millennium Year 2000.

In the meantime, David Crighton (1942--2000), who had been elected to the Chair of Applied Mathematics in 1986, was taking on increasing burdens of responsibility. He succeeded me as Head of DAMTP in 1991, when my term of office came to an end, and he gradually took over editorial responsibility for *JFM*, and the Chairmanship of EUROMECH (the European Society for Mechanics), which had been founded by Batchelor and Dietrich Küchemann in 1964.

Tallahasee 1991

Relieved from Head-of-Department responsibilities, I was able to take up some work with David Loper at the University of Tallahassee, Florida, on research that evolved into what I later described as 'magnetostrophic turbulence', the conjectured state of flow in the deep liquid core of the Earth. Linty and I spent an enjoyable fortnight in Tallahassee in March 1991, during which Dave took us on a day trip to his family home on St George Island, where we enjoyed a sumptuous feast of clam chowder and other delights.

Later that year, I was able to air the beginnings of our approach to the geomagnetic dynamo process in my Union lecture *"The Earth's magnetism; past achievements and future challenges"* at the Congress of the International Union of Geodesy and Geophysics (IUGG) in Vienna. Unfortunately, the microphone system was inadequate, and there was so much background noise that only those in the front rows of the large lecture hall could hear what I had to say! I was encouraged that Bertha Jeffreys (1903--1999) and James Lighthill were in the front row; Bertha kindly told me after the lecture that she wished she had been my student, a compliment that I cherished.

With Bertha Jeffreys in Cambridge c.1991

KITP Santa Barbara, 1991

Following the 1989 IUTAM Symposium on *Topological Fluid Mechanics*, John Greene (1928--2007), of General Atomics and

UCSD, was so enthused by the theoretical problems that this subject raised that he recommended a programme to cover these problems at KITP, Santa Barbara. This was held for the four-month period August to December 1991; it served to train me for the running of the INI programme on dynamo theory which was to be held just one year later, July to December 1992. There were several participants at the KITP programme from the USSR, which was in turmoil throughout that period, culminating in the resignation of President Gorbachev and dissolution of the USSR on 26 December 1991. Among these Soviet participants was Sasha Ruzmaikin, who had translated my monograph into Russian in the late 1970s. He was super-excited about the dramatic political situation in USSR, and he resolved to emigrate to USA as soon as he could, following this programme.

INI Dynamo theory 1992

Prince Philip, Duke of Edinburgh and Chancellor of the University of Cambridge, at the formal opening of INI in October 1992.
From left, HKM, Andrew Soward, Uriel Frisch,
Prince Philip, and Sir Michael Atiyah, founding Director of INI.

The Isaac Newton Institute for Mathematical Sciences (INI) opened on 3rd July 1992, with the simultaneous opening of two 6-month programmes, one on dynamo theory, for which I was the principal organiser, the other on low dimensional topology and quantum field theory, organised by Raymond Lickorish of the Cambridge Department of Pure Mathematics and Mathematical Statistics (DPMMS). Four lectures were given before an invited audience on the opening day; Vaughan Jones [1952--2020] gave one of these; he had won the 1990 Fields medal for his work leading to discovery of a new invariant (the Jones polynomial) for knots and links. I was privileged to give another. We were instructed to give a one-word title for our lectures. Vaughan Jones chose *Knots* for his. I chose *Flow* for mine; this gave me the opportunity to quote the ancient statement πάντα ρει [everything flows], attributed to Heraclitus in the sixth century BC. From then on, it was plain sailing!

The more formal opening of INI by Prince Philip, Chancellor of the University, took place on 30th October 1992. The photo shows him with Sir Michael Atiyah, chatting with my co-organisers of the dynamo theory programme, Andrew Soward and Uriel Frisch. We had shown him the experiment that shows the formation of a cusp on the surface of a viscous liquid, forced to converge by counter-rotation of two submerged cylinders. This was an investigation that I had carried out with a recent Korean visitor to DAMTP, JaeTak Jeong. Prince Philip's comment about the formation of the cusp was *"That's pretty obvious, isn't it?"* But the mathematics required to explain this phenomenon was far from obvious!

Kyoto 1993

At the invitation of Shigeo Kida, I spent three months at RIMS Kyoto, accompanied by Linty, from January to April 1993. Actually, with

Japanese precision, the visit had to be for 90 days, neither more nor less. We therefore broke our flight out for a three-day stay in Bangkok, so that we arrived in Kyoto refreshed and recovered from jet-lag, i.e. ready to start work immediately. We stayed in an apartment at the Shūgakuin Visitor Hostel, from where I could get to RIMS by train to the station Demachiyanagi (what wonderful names!) and a short walk; or by bike when the weather permitted. It snowed gently every morning and thawed every afternoon, until the sudden advent of spring and the famed cherry blossom in early April.

We were happy to renew acquaintance with old friends, Tomomasa Tatsumi, veteran of turbulence research, and the Yosinobu's who had spent the academic year 1961/62 in Cambridge. During this period, I was invited by Masaru Kiya to give a seminar at Hokkaido University in Sapporo, where we saw the magnificent ice sculptures for which Sapporo is famous. Later, I gave seminars also at Fukuoka University in Kyushu, travelling there from Kyoto by the bullet train (Shinkansen), and at the University of Tokyo at the invitation of Tsutomu Kambe.

My visit was scientifically fruitful. Together with Shigeo and with Koji Okhitani, we completed a paper *"Stretched vortices - the sinews of turbulence"*, again in precisely 90 days; I submitted this to *JFM* on our return to the UK, and it was published in 1994.

With Tomomasa Tatsumi in Kyoto

Towards the end of our stay in Kyoto, I was invited to contribute an informal article to the RIMS Newsletter, about any aspect of our visit; here is what I wrote:

Croquet in Kyoto

"Down Higashioji Dori, just after an Establishment with the mysterious name *"Magnoria: Coffee Stained Glass"*, a road slants off to the left towards Chayama and Kyoto University. I cycle down there each morning observing as well as I can the rules (or non-rules!) that pertain to cycling in Kyoto. On a particular cold crisp morning last week -- it was late in March and early in the morning -- I came upon a little corner of Paradise: a white sanded area on the right where children were playing with their toys and a group of elderly citizens were playing a game called CROQUET. Nothing, you may say, so extraordinary about this; but what is extraordinary to my mind is that Croquet is a game that is quintessentially English (despite the French pronunciation, like CROAKEY, rhyming with jokey). I have never found this game in France, or in Germany, or in Italy. You find it only, among Western countries, in England, usually on the well-tended lawns of stately homes or Oxbridge Colleges. It is a game played with wooden balls, mallets and hoops. There are complicated rules of procedure, but basically two teams have to propel their balls through the hoops in prescribed order towards a finishing post, doing their best to inhibit the progress of their opponents in the process, the first team to complete the circuit being the winner. There's a chapter in *Alice in Wonderland*, in which this game is played by Alice in dreamlike confusion. You can imagine how surprised I was to find this same game being played here in a back-street of Kyoto. I parked my bike and stayed

Croquet in Kyoto

awhile to watch the unfolding drama. A campfire by the side of the court provided warmth for the umpire who invited me to join him to observe and enjoy the play. The rules were different from those of the game I knew in England, but the general objective was clearly the same: to defeat the opposing team by any combination of skill with the mallet and guile of the mind; the laws of impact dynamics are naturally relevant!

"In England, a degree of confrontation is evident from the outset in such contests; each team immediately endeavours to blast the balls of the opposing team by 'roquet' and 'cannon' to the furthest corners of the court, and thus to render them impotent in the subsequent struggle for superiority. Here in Japan, I observed a different strategy: the two teams avoided direct confrontation, each preferring to establish a strong position through delicate manoeuvring, in different corners of the court. Of course, confrontation was in the end inevitable, and when it came it was fierce and destructive; the red team triumphed over the white, and then all participants assembled round the campfire, where the umpire had by now prepared refreshments -- sweet cakes and green tea.

"Later that same day, I caught my first glimpse of cherry blossom in full bloom and of parasols (as opposed to parapluies!) protecting their kimono'd holders from the warming sun. Many such vivid and colourful impressions of Kyoto life will remain with me from my three brief months in this enchanting city; but, in a curious way, it is the croquet game, with its peculiarly English associations, that has given me a particular affection for the gentle and exceptionally friendly people who live in these little backstreets between the Eizan Railway and the Takano River."

Cargèse by car, 1993

In August 1993, I took part in the Summer School "*Turbulence, Forte et Faible*", organised by Patrick Tabeling at the Institut d'Etudes Scientifiques de Cargèse, in Corsica. Linty and I took our car by train from Calais to Nice, then by sea to Bastia; we then drove across Corsica via Calacuccia, with occasional sight of wild boar on the way, to Porto, then on a hair-raising road down the coast to Cargèse. Russell Kulsrud was there, delightfully entertaining over the outdoor lunches in that stunning environment; also Martin Kruskal (1925--2005, "one of the most versatile theoretical physicists of his generation", according to the *Biographical Records of the Royal Society*, of which he was a foreign member). Their presence ensured that this was indeed a very lively Summer School.

Ecole Polytechnique, Palaiseau

Patrick Huerre, Director of LadHyX [the Laboratoire d'Hydrodynamique at the École Polytechnique in Palaiseau ("X")] asked me whether I might be able to teach a course at X, as a "Professeur à temps partiel", i.e. in a part-time capacity. On my acquiescence, he put this proposal to the relevant committee at X, which naturally scrutinised my cv with a fine-tooth comb. Patrick told me afterwards that one committee member had observed that, since I had spent a term at the Lycée Henri-IV as an 18-year-old schoolboy, I should without doubt be appointed to this position! It seemed that this comment swayed an otherwise sceptical committee, and so it came about that I was appointed and served 7 years in all at X, 1993-1999, preparing and giving two courses of lectures ('majeures'), one on '*Microhydrodynamique*' and the other on '*Tourbillons et Turbulence*'. For two of these years, I was able to spend three months in Paris with Linty, the second of these in a fifth-floor rented apartment in Rue du Pot de Fer, just off the Rue Mouffetard in the Quartier Latin. From there, three times a week I would take an early-morning RER from Luxembourg to Lozère in the suburbs, then after a 'grand cafe au lait' climb the ~140 steps to the plateau on which X is situated (it is said that *les Professeurs* retain their positions at X only for so long as they can climb these steps!).

For each of the other five years, I commuted from Cambridge to Paris for two days each week for the two-month duration of the courses. The lectures were arranged to fit this schedule, and to fit my continuing responsibilities in Cambridge. I am proud that the written versions are retained in a section of the library of X and sit alongside the historic courses of Ampère, Cauchy, Navier, and other such illustrious 19th century figures. I was given to believe that I was the first non-French Professeur at X, at a time when efforts were being made to open the École to a more international scientific community. On my resignation in 1999, I was presented with a ceremonial sword of the École Polytechnique, a signal honour, and a useful disciplinary symbol in my role as Director of the Isaac Newton Institute back in Cambridge!

The École Polytechnique was originally founded as a military academy under Napoleon Bonaparte, and it is still supervised by the French Ministry of Armed Forces. On the rare occasions when I had to present myself in the office of *'mon Général'*. I couldn't help noticing the sumptuous quality of the carpets and furnishings in this administrative section of the École.

The military discipline was brought home to me when, on one occasion I was stricken with pneumonia one Monday evening while at our apartment in the Rue du Pot de Fer, having lectured at X that morning. I was taken by ambulance down to the Hôpital Pitié-Salpêtrière, the same hospital where Princess Diana was to die so tragically in 1997; there, with Linty by my side. I lay for some hours on a stretcher, alongside a number of drunks brought in from surrounding streets, until finally given an injection which brought my temperature under control, and enabled my return to our apartment the next morning. I was due to lecture at X the following day. Fortunately Robert Sadourny was able to stand in for me on that day and also to drive me to X two days later for the next lecture: you just weren't allowed to miss lectures, no matter how ill you might be. So on the Friday that same week, I was propped up against the lectern, and I gave my 90-minute lecture, pumped full of paracetamol --- whence our family joke "when you get to Paris, eat 'em all". Mercifully, I survived, the students being very understanding! Linty was very relieved when I emerged from the lecture room, still alive though hardly kicking.

SEDI at Whistler BC, 1994

In August 1994, Linty and I were in British Columbia following a visit to Linty's twin brother Bobby, now settled as the Presbyterian minister of Comox on Vancouver Island. We hired a car in Vancouver and drove up to Whistler in the dark and in a downpour on a treacherous mountain road to attend the inaugural meeting of SEDI (*Study of the Earth's Deep Interior*), where Dave Loper and I were able to present our work as it was then developing. This work was taken up by Martin St. Pierre who showed that our compact 'rising blobs' were unstable to a 'slicing instability' -- becoming like a (very large) loaf of sliced bread -- which led us to develop a less constrained model of 'magnetostrophic turbulence' --- evidently the best thing since sliced bread!

The next two years were punctuated with visits to Madrid, at the invitation of Javier Jimenez; to Hong Kong, at the invitation of

Volodya Vladimirov, who had a position as Professor at HKUST (the Hong Kong University of Science and Technology); and to Kittila (in Finnish Lapland) for a Nordic Winter School, at the invitation of Axel Brandenberg. This last was in February 1996; we hoped to see the Northern Lights, but alas there was too much cloud. Eugene Parker was at this meeting, and we had a good discussion concerning the formation of current sheets, on which he had recently published his book "*Spontaneous current sheets in magnetic fields*".

Gaulrig 1996

Later in 1996, I took my mother Emmeline on a trip to Speyside, the land of her Grant ancestors, where we located the 1820 gravestone of Isobel Grant in the Kirkmichael graveyard just north of Tomintoul. We also located the ruins of Gaulrig, where this Isobel had met her tragic death in the illicit still. Emmeline met her death equally tragically six months later (March 1997) as the result of a car accident on Barton Road just outside Cambridge. She was then 88 years old. She had received a visit from the Presbyterian minister of St Columba's church the previous evening at her home in Grantchester. He had noticed her Honda at the door and expressed surprise that she was still driving. She replied that it was her lifeline and that "*when it goes, I go*". This was sadly to be sooner than she had anticipated.

Director of INI 1996--2001

By this time, I had succeeded Michael Atiyah as Director of the Isaac Newton Institute. I was appointed for the five-year term 1996--2001. During my tenure, I continued to hold my Professorship in DAMTP (though relieved of some lecturing duties) and the University continued to pay my salary.

My main concern in October 1996 concerned the finances of the Institute, and I had bargained with the University over the preceding months (as a condition of my accepting appointment) to provide some limited financial guarantees. The Scientific Programmes of the Institute were running well, and some good proposals for future programmes were under consideration by the Scientific Steering Committee. The major problem was to provide adequate funding for these programmes, and to establish a sound financial platform for the longer-term future.

Sir Michael had approached Lord (Victor) Rothschild, who was an Honorary Fellow of Trinity College, for support for the Institute when it was still at an early planning stage. This led to a donation of £250k to support a succession of 'Visiting Rothschild Professors' to the Institute, a donation made on behalf of NM Rothschild & Sons by its Chairman, Sir Evelyn de Rothschild, in 1991 following Victor's death in 1990.

My first step was to appeal to the Isaac Newton Trust, which had provided support during the initial years. In January 1997, the Trust agreed to provide an interest-free loan of £1m, this loan to be convertible to endowment, subject to the raising of a matching sum from external sources. This provided invaluable leverage in subsequent approaches that I made later that year.

In particular, I approached NM Rothschild & Sons in the hope that they would at the least be prepared to extend their support of the visiting Professorship scheme, which had been very successful. By lucky chance, 1997 happened to be the 200th anniversary of the arrival of the original Rothschild (Nathan Meyer) in England, and the bank wanted to mark this bicentenary in special ways. In the end, Sir Evelyn agreed to donate £1.75m towards the endowment of a Rothschild Professorship, this Chair to be held in conjunction with the Directorship of the Newton Institute. This was a marvellous breakthrough, which immediately met the condition of the Newton Trust and provided some security for the future. The endowment, announced in May 1998, was built up over the subsequent 4-year period, being completed in time for the appointment of my successor Sir John Kingman FRS in 2001, the first N.M.Rothschild & Sons Professor of the University.

With Dill Faulkes, when we celebrated his generous donations to Cambridge University

One of our participants in the programme *"Mathematics of Atmosphere and Ocean Dynamics"* (July to December 1996) was Professor Ian Roulstone (University of Reading). He was conscious of my need to raise funds, and kindly introduced me to his bell-ringing friend Dill Faulkes. I invited Dill to Cambridge early in 1997, and struck up a friendship with him. He was indeed interested in the work of the Newton Institute, and said that he might be able to help, although not for some time. I kept him informed about all new developments at the Institute, and it was early in 1999 that Dill made his wonderfully generous donation of £1m to the Institute. He was keen that this should be used for 'bricks and mortar', and we therefore used the donation for the construction of the Faulkes Gatehouse, providing much needed expansion space. This building was completed and ready for use in 2001 (by which time Dill had, through his continuing relationship with the Newton Institute, made a further munificent donation of £2.5m to the University towards construction of the Geometry Pavilion on the main mathematics (CMS) site.

A considerable boost to our fund-raising activity at the Newton Institute was provided by the award of a Queen's Anniversary Prize in 1998. This was the third round of these prizes, and the first occasion on which a submission had been made with the approval and on behalf of the University of Cambridge. The prize was awarded at a ceremony at Buckingham Palace in February 1999. I had been asked to bring seven students with me to Buckingham Palace for this occasion. The INI has no students, only visiting researchers including many post-docs. I chose seven post-docs from seven different countries, in order to emphasise the international scope of our activities. When The Queen did her walk-about chatting to many who were present for the reception that followed the formal ceremony, she asked one of my post-docs who was working in cosmology, her stock question *"and what do you do"* to which he replied *"I study the origins of the Universe, Ma'am".* I think The Queen was quite non-plussed as she responded *"Oh, how very interesting"* and moved on.

Her Majesty Queen Elizabeth awards the Queen's Anniversary Prize, 1998. I was accompanied by Sir Alex Broers, then Vice Chancellor of Cambridge University

Following review by the Engineering and Physical Sciences Research Council (EPSRC) in 1997 (chaired, as chance would have it, by Sir John Kingman, at that time Vice Chancellor of Bristol University) the rolling grant of the Institute was extended to February 2002. During 1998, we engaged in intensive follow-up discussion with EPSRC concerning future modes of funding beyond 2002, including the possibility, favoured for a time by EPSRC, of funding on a programme-by-programme basis. This would have posed very grave problems for the Institute which requires quite a long-term funding horizon in its planning, but fortunately our arguments for an adequate funding horizon prevailed, and in April 1999 EPSRC agreed to extend the rolling grant to February 2008, subject to triennial review in 2002 and 2005. This most welcome news paved the way for the appointment of Sir John Kingman as my successor in office, more than covering the period of his tenure 2001-2006.

In 1999, we undertook the design of twelve *Posters in the Underground* as our contribution to World Mathematical Year 2000. This project was funded by EPSRC's *Public Awareness Initiative*. The posters appeared month-by-month during 2000 in trains of the London Underground. They proved very popular, and were subsequently widely distributed to schools and Universities, not only in the UK, but also in South-East Asia, Australia and USA (in slightly modified form).

One of the monthly *Posters in the Underground* © Andrew Burbanks.

In 2000, I approached Robin Fleming [1933--2020], Director of the Robert Fleming Investment Trust, and a 3rd cousin of my wife, Linty. This led to a most welcome donation of £240k from the PF Charitable Trust, spread over 4 years, to be used preferentially but not exclusively for the support of participants from Scotland in Institute programmes.

We also received a generous bequest (amounting ultimately to just over £1m) from a donor in USA who wished to remain anonymous. He had corresponded with me about 18 months before his death, concerning the possibility of making such a bequest. His will specified that the residue of his estate should be divided equally between the Royal Society and the Newton Institute, which I took to be a fitting tribute to the Newton Institute during my final year of office!

Twenty-eight scientific programmes were held during the period of my Directorship, spanning the physical, biological and engineering sciences. These attracted a total of more than 1000 long-stay participants and more than 5000 additional short-stay participants in workshops held within these programmes.

I was fortunate to have two outstanding Deputy Directors during my five years at the Institute: Noah Linden for the first three, Robert Hunt for the last two. They were an unfailing source of help and moral support. I was also fortunate to have a wonderfully committed and loyal staff who ensured the smooth operation of the Institute in all its wide-ranging activities.

My final achievement in the old millennium was to have a boules court established at the Newton Institute, accessible through the french windows that open from the main circulation area. I felt this would provide welcome relaxation for our international visitors, after the long hours of intense mathematical effort that they would habitually spend in their offices or in lively discussions at the blackboards. The existence of this boules court gave me an excuse to write a poem two years later in response to a directive from Hassan Aref. This poem, which celebrates the game of boules in appropriately tongue-in-cheek manner, is reproduced in Chapter 10, together with an explanatory introduction.

Skreen, Ireland 1998, Sir George Gabriel Stokes

Linty, turned to stone, as we searched for the grave of George Gabriel Stokes, in the Mill Road cemetary

In August 1998, I drove across Ireland with Linty to attend a meeting in Skreen, County Sligo, the birthplace of Sir George Gabriel Stokes (1819--1903), whose seminal work in the 1840s led to the 'modern' derivation of the Navier-Stokes equations. It is said that Lord Kelvin stood at the graveside of Stokes in 1903 declaring *"Stokes is dead; I shall come to Cambridge no more"*. Their wonderfully revealing scientific correspondence had spanned the previous 50 years.

The meeting in Skreen was deliberately held in the primary school that Stokes had attended as a child, with only chalk and a small blackboard on an easel to aid the delivery of our lectures. A group of about 20 UK fluid dynamicists squeezed into the classroom, and enjoyed each other's presentations. Sir Michael Berry and Howell Peregrine from Bristol were among those present, also Alan Newell from the University of Arizona. It was a most unusual meeting, but a great way to commemorate a towering figure in the history of fluid mechanics.

Humphry Davy Lecture, 1998

The Humphry Davy lecture is one of the prize lectures of the Royal Society, and is given in alternate years at the Royal Society and at the Académie des Sciences in Paris. In 1998, it was the turn of Paris and I was the chosen victim. I lectured at the Académie before a venerable audience of Academicians on *"Energy minimisation under topological constraints"*. But what particularly pleased me was to see M. and Mme. Gama there, the extraordinarily hospitable couple who had welcomed me forty-five years before to their home in the Bld Raspail during my 1953 'stage' at the Lycée Henri IV. Their sparkling daughter Annie, a mathematics teacher in Paris, who as a child had teased me mercilessly, was with them also. It was wonderful to see them again in such a magnificent setting "under the Coupole" after all these years.

CHAPTER 7
The 00s: Toying with Spin

George Batchelor died on 30 March 2000, a sad day for DAMTP, the department that he had founded in 1959, and to which he had been devoted for 40 years. Parkinson's disease finally got the better of him, although he had continued to struggle against it throughout the 90s. In my *Biographical Memoir* for the Royal Society, I wrote in summary *"Batchelor was ascetic by instinct, frugal without meanness, with a dry humour, and perhaps giving an impression of austerity in later years, although always in practice approachable and a source of sound advice to colleagues and research students alike. He will be remembered as a man of great scientific integrity, penetrating judgement and deeply held convictions."*

David Crighton died two weeks later, aged 57, a tragedy for DAMTP, in which he had succeeded me as Head of Department in 1991. David had taken on the Mastership of Jesus College in 1997, but was diagnosed with cancer soon after that. If anything, this only intensified his focus on managerial and editorial responsibilities, which he pursued to the awe and growing esteem of his colleagues to the bitter end. He was himself gravely ill in Addenbrooke's Hospital when he received the news of Batchelor's death, and he immediately dictated his warm tribute to George from his hospital bed. He died on 12th April, my 65th birthday, which made this all the more poignant for me personally. Another sad, sad day.

President of IUTAM 2000-2004

At ICTAM Chicago in August 2000 I was elected President of the International Union of Theoretical and Applied Mechanics (IUTAM). I have devoted Chapter 1 to IUTAM matters that have concerned me over the whole period 1960-2024. I should just record here my debt to George Batchelor for having introduced me to IUTAM in the first place, while still a research student; and my deep friendship with David Crighton who told me in January 2000 how pleased he was to learn that I was running for President of this esteemed international organisation. During his final year as Master of Jesus, I called on him every week at the Master's Lodge to deliver a loaf of my home-baked wholemeal bread in the hope that this would help him in his battle against cancer; it was the least I could do in a rather hopeless situation.

In 2001, I was awarded an Honorary D.Sc. at my *alma mater* Edinburgh University. The photo shows my sister Lindesay arriving in the Old Quad for this ceremony, as always in the pink.

Euler's disc

In 1998, I had purchased a toy, advertised as 'Euler's disc', intending this as a Christmas present for my grandchildren. I found the toy so intriguing that I am embarrassed to admit that I kept it for myself! What

My sister Lindesay arriving at the Old Quad in Edinburgh for the ceremony of the award of my D.Sc

was intriguing was that it appeared to exhibit a 'finite-time singularity', that is, a property that becomes infinite within a finite time. Euler's disc is a fairly heavy disc that can roll and rotate on its edge on a smooth plane or on a slightly concave surface. As it does so, it slowly loses energy and settles on the 'substrate' in a finite time (usually a few minutes), but the precessional angular velocity appears to increase without limit during this finite time. In the final stage of settling, the air between the disc and the substrate is rapidly sheared, and thus dissipates at least some of the energy of the motion. My analysis of this mechanism was published as a Brief Communication in *Nature* in 2000. It immediately provoked dissenting opinions, concerning the true mechanism of energy dissipation, and a flurry of experiments indicating that 'rolling friction' might in fact dominate. A lively debate ensued!

Euler's disc at the Newton Institute

The rising egg problem

I gave a colloquium on Euler's disc in DAMTP in 2001, and in passing I mentioned the 'rising egg problem': if you spin a hard-boiled egg on a smooth table (make sure it's hard-boiled!), it will rise on one end and spin for some time in this vertical position until it finally wobbles and falls over. I rashly stated that it should be easy to explain this phenomenon. A week later, Yutaka Shimomura, a visitor from Japan who had attended my lecture, came to my office with a strained look, and said he had been trying all week to solve this dynamical problem without success -- and I had said that it was easy! We proceeded to work on it together. It was indeed much more difficult than I had so rashly predicted.

A month or two later, I was still struggling over the problem while travelling to Turin by train to receive the Pannetti-Ferrari prize, which I hardly deserved! But then suddenly the light dawned, and I found a way through the morass of equations governing the spinning-egg behaviour. Yutaka and I completed a paper which appeared as a further Brief Communication "*Spinning eggs -- a paradox resolved*" to which *Nature* added the timely sub-text "*An explanation for an odd egg performance is rolled out in time for Easter*". My research student Michal Branicki joined in our subsequent work that provided a generalised theory for the spinning, rolling and slipping behaviour of such axisymmetric bodies. What was important was to identify the property that remained approximately constant (not angular momentum in this situation) and to minimise the total energy subject to this constraint.

The rattleback

At the same time, together with Tadashi Tokieda, I got involved in the related problem of the rattleback (or 'celt'), a canoe-shaped object which exhibits a paradoxical behaviour, in that it prefers to spin in one direction rather than the other; in fact, if you try to spin it in the 'wrong' direction, it wobbles, as if

At the Royal Society final evening reception: HKM, Tadashi Tokieda, Konrad Bajer and Michal Branicki

quite upset, and reverses direction. The cosmologist Hermann Bondi (1919--2005), who had worked with Fred Hoyle in promoting the ill-fated steady-state theory of the Universe, had studied the stability of the steady rotation of this object, but had not been able to explain the reversals. Tadashi and I managed to explain these by means of a 'two-timing' analysis; the behaviour was a good example of the 'breaking of reflection symmetry', a key to the understanding of the generic cosmic process of magnetic dynamo action. Mechanical toys like this are really interesting when they can act as prototypes of processes that occur in geophysical or astrophysical contexts.

Around this time, Joe Keller from Stanford visited us at our home in Cambridge. Here you see him demonstrating how gravity can defeat friction if things move fast enough. You place a record sleeve (the old-fashioned kind for 33rpm records) on a glass of water, and balance an egg (not necessarily hard-boiled this time) on this on the sleeve of a match box, suitably deformed. Then if you strike the record sleeve away horizontally and sufficiently rapidly, as Joe has done here, the egg will drop into the water without breaking. The effect is dramatic, but requires confidence. Any hesitation will result in catastrophe!

Joe Keller at our home in Cambridge

I proposed *"The Dynamics of Spin"* as an exhibit for the Royal Society Exhibition in 2007, for which Andy Burbanks designed the posters. We then took the exhibit to the Mumbai TechFest in January 2008, where the toys naturally attracted the attention of huge numbers of young enthusiasts. I went on from there to give a seminar at the Indian Institute of Science in Bangalore, where I was very happy to have renewed discussions with Roddam Narasimha, who had served on the Bureau of IUTAM during my time as President, and to learn of the research at this outstanding Institute.

My team preparing for work at TechFest, Mumbai, Jan. 2008; Yuri Sobral, Andy Burbanks, Michal Branicki, Kiran Singh and myself

Kiran demonstrates a mechanical toy to a group of young enthusiasts

An unsuccessful bid for funding

John Coates (1945--2022), who had been Head of DPMMS in the 1990s, introduced me to John Fry, Director of *Fry's Electronics*, a 'big-box' store chain in the United States. He had an interest in mathematics, and we welcomed him on a visit to the Newton Institute. John Fry was extremely rich, and was naturally identified as a possible benefactor. I took him and his wife to dinner in Trinity the evening before their departure for Texas. In the Combination Room after dinner, one of my colleagues asked Fry "When does your flight take off?" "When we get to the airport" he replied.

I visited John Fry at his home in Palo Alto some months later. He took me on a golf-cart tour of his private golf course that he was proud to have designed. The greens were immaculate, and the bunkers finely raked, but it was all rather strange because not a single golfer was to be seen; it was a billionaire's playground! John quizzed me that evening about our plans for the Newton Institute, but alas, nothing came of his welcome curiosity. I note that Fry's Electronics went bankrupt in 2021, so perhaps in retrospect this should not be seen as a misfortune.

The Zakopane limericks, 2001

Konrad Bajer organised an IUTAM Symposium on the subject *"Tubes, Sheets and Singularities in Fluid Dynamics"*. This was held in September 2001 in Zakopane, to which I was very happy to return, as I recalled with nostalgia my first visit there in 1963, so many years before. I made it my project to summarise each lecture of the Symposium in the form of a limerick, as much for my own benefit in trying to extract the essence of the lectures, as for the subsequent amusement of the participants. This for example is my summary of the presentation of Renzo Ricca, which I am sure he will forgive me for reproducing: *"Topological arguments show/ that the energy's bounded below;/ but what's so engrossing's/ the number of crossings,/ from which my new insights will flow."* Most of the limericks were actually published in the Proceedings of the Symposium; I suspect this is a unique achievement in the annals of science!

Retirement 2002

I retired as Professor of Mathematical Physics in 2002 at the then compulsory retiring age of 67. I gave my final course of lectures on fluid mechanics in the Michaelmas Term of my final year. Again, I decided to summarise each lecture in a limerick, which I was confident students would remember if all else was forgotten. This sort of thing: *Profiles of speed parabolic/are common in channels hydraulic,/ in pumping of blood,/in sliding of mud,/and in suction of drinks alcoholic.* I have been happy to meet former students years later, who assure me that they remember the limericks, but little else. So education does serve a purpose.

Iran, May 2002

In 2002, I received an invitation to lecture at the Asian Congress in Fluid Mechanics (ACFM), held that year in Isfahan, Iran. It was a wonderful opportunity to visit this cradle of civilisation. As President of IUTAM, I could hardly decline such an invitation, and I was indeed very glad to accept, although conscious that the atmosphere in Iran might still be tense so soon after the 9/11 twin-towers disaster. And so it was! At the opening of the Congress, two exceedingly solemn bearded men in black came onto the stage and uttered some words in Arabic from the Quran. I can't say this had the uplifting effect that was perhaps intended, but the scientific presentations soon took over. Hans Hornung from CalTec gave the opening lecture; then Paul Linden on building ventilation, very apposite for Isfahan. I gave the Zhou PeiYuan lecture. We were served refreshments in a pleasant courtyard by attentive hosts, particularly the charming women students in the alluring chadors.

Osama Sanu welcomed by hookah smokers under one of the arches of Si-o-se-pol

The sights of Isfahan are fabulous, but what I remember most was sharing a hookah with an animated group of students under the arches of Si-o-se-pol ('bridge of thirty-three spans') over Zayanderud, the river that runs through the city.

The post-Congress tour was to Shiraz, for which only 8 of us had enlisted; among them Min Chong and Anne Hellstedt from Melbourne University. Our guide, Jamshid, took us by mini-bus to Persepolis, then to the Pasugardae World Heritage site, and Naqsh e-Rostam, a day on which I encountered 2500 years of history.

After a day in Shiraz visiting the tomb of the poet Havez, who wrote this sobering thought for monarchs (here as translated by Gertrude Bell):

Party of 8 in Persepolis

"The Sultan's crown, with priceless jewels set,
Encircles fear of death and constant dread;
It is a head-dress much desired — and yet
Art sure 'tis worth the danger to the head?"

--- yes, after this day, I had one more day when I was guided by taxi-driver Mansour through the Zagros Mountains and by the descent, hot and hotter yet, to Bishapur, where the Anahita Temple was as incredible as it was ancient.

Blaise Pascal Chair 2002/2003

Tim Pedley, who succeeded David Crighton as Head of DAMTP, drew my attention on my retirement to the Blaise Pascal Chair, tenable in the Île-de-France, i.e. the greater Paris region of France. Blaise Pascal [1623--1662], he who stated (as relevant in the current era of fake news as in his time) *"Truth is so obscure in these times, and falsehood so established, that, unless we love the truth, we cannot know it."* I applied for this Chair, and was very fortunate to be appointed for 12 months that could be spread in any way over a a two-year period. I elected to hold it at the Ecole Normale Supérieure (ENS) in Paris. I gave a course of lectures on *Topological Fluid Dynamics* at the Institut Henri Poincaré in 2003 and numerous seminars at many of the Paris Universities.

One of these seminars was in the series organised by Vladimir (Dima) Arnol'd in what was then the Denis Diderot University (Paris VII). Arnol'd was by then dividing his time equally between Paris and Moscow. His seminar series was a great attraction for researchers from all over Paris and beyond. He chaired the seminar that I gave, which is to say that, after my opening remarks, he leapt to his feet and took over at the blackboard for the next 10 minutes before relinquishing the floor to me. I took this to indicate his enthusiasm for the topic that I was proposing to discuss. He was in an extraordinarily excitable mood, but as always, his comments were of a most penetrating character.

We were able to rent again the now familiar apartment in Rue du Pot de Fer, just round the corner from ENS, for the duration of our stay in Paris. This was made even more pleasurable by the fact that our elder daughter Hester, her husband Fred Tingey, and their four young children Chloe, Tabitha, Alfie and Bathsheba, were also in Paris for much of the time, actually in Maisons Laffitte, an affluent north-western suburb of Paris. Freddie worked for BNP Paribas, and had been seconded to Paris for much of the period that we were there. Linty and I were able to visit the family from time to time in Maisons Laffitte, and equally they were able to visit us in Rue du Pot de Fer.

Hester was much involved in horse riding with the children during these two years, and they got so attached to two ageing horses, Sifonette and Prince, that Hester had them shipped back to England at the end of their secondment (they being otherwise destined for the knackers' yard). They gave great pleasure to the family back in England for at least the next 10 years.

Hester with Chloe and Tabby, horse-mad in Maisons Laffitte

While we were in Paris, a kind neighbour at Rue du Pot de Fer introduced us to her friend Marjane Satrapi, who had just published her brilliant graphic novel *Persepolis*, describing her childhood in Iran during and after the 1979 Islamic revolution. We visited Marjane at her studio near the Place de la Bastille. From there, we soon gravitated to a bar together with Marjane for tea, rapidly evolving to several glasses of cognac --- it was a very convivial occasion.

Yves Couder and bouncing droplets

In January 2003, I met with Yves Couder (1941--2019) and three of his students, to discuss the 'bouncing drop' phenomenon. When a droplet of silicon oil is released on a dish of the same fluid which is vibrated up and down at a suitable amplitude and frequency, the droplet is not absorbed by the underlying fluid, but instead bounces repeatedly from the surface, a behaviour that continues more or less indefinitely. This was the beginning of Couder's great discovery that such bouncing droplets generally move on the liquid surface, and in so doing exhibit 'wave-particle' behaviour like the mysterious dual behaviour of electrons and other elementary particles as revealed by quantum mechanics. The bouncing particle in effect interacts with the wave field that it creates, and its behaviour can only be understood in terms of this wave-particle duality. This wonderful discovery opened the way to understanding quantum mechanical effects through a classical fluid-dynamical model. I believe that it was largely on the basis of this discovery that Yves was elected to the Académie des Sciences in 2013. Tragically, he died of cancer in 2019, when the significance of this seminal work, greatly promoted by John Bush (MIT) and others, was gaining ever-increasing recognition in the worldwide scientific community.

The African Institute for Mathematical Sciences (AIMS)

I became involved in the AIMS initiative in 2001, when I was still Director of the Newton Institute in Cambridge. Neil Turok sought my advice concerning the setting up of an Institute in South Africa similar to the Isaac Newton Institute in Cambridge. I still remember his phone-call. My immediate reaction was: what a brilliant idea! But the difficulties seemed formidable, and in particular the Newton Institute model of a 'visitor research institute' did not seem entirely appropriate to the situation in Africa. Here the most pressing need at

Neil Turok, Founder of AIMS

that time appeared to be to provide a diploma-style course at graduate level to bridge the gap between the shifting sands of undergraduate attainment and the frontier at which research problems might be confidently addressed. A crash course of highly innovative structure was needed, something like the well-established Part III mathematics course in Cambridge, but adapted to Africa's most compelling needs. Neil's vision might evolve naturally over a period of years from the foundations laid by such a course. I enthusiastically supported this concept, and encouraged Neil to give initial priority to graduate-level education, and to seek to engage lecturers from the international community as well as from neighbouring South African universities, to provide the style of teaching that this diploma course would require.

I play my part in discussions at the 2004 AIMS workshop

The AIMS initiative was timely in relation to the priority that was increasingly being accorded during the 1990s to 'capacity-building' by international organisations, and particularly by ICSU (the International Council for Science), whose *raison d'être* is the *"strengthening of international science for the benefit of society"*. ICSU is the umbrella organisation for at least 30 International Scientific Unions, including the International Mathematical Union (IMU). As President of IUTAM, I had an opportunity to promote the AIMS initiative at the ICSU level. On behalf of IUTAM, and with the support of five other International Unions (IUGG, IMU, IUPAP, IUPAC and IAU), I applied in February 2003 for a capacity-building grant to help with the establishment and promotion of AIMS during its crucial first year 2003/4. A grant of $100,000 was awarded, enabling us to hold a capacity-building workshop at AIMS in April 2004, hosting representatives from 20 African States; this served to promote the distinctive character of AIMS, and it was in fact here that Neil Turok's vision of a network of AIMS centres based on the AIMS-South Africa prototype was first aired and discussed. Originally called AMI-NET, this network evolved into AIMS-NEI (the AIMS Next-Einstein Initiative), with further AIMS centres established in Senegal, Ghana, Cameroon, and Rwanda where the administrative centre of AIMS is now located.

Muizenberg, South Africa

The original AIMS-South Africa conjured up for me a wonderful place and a wonderful vision. The place in perhaps obvious ways -- that beautiful Muizenberg beach with its rolling surf and rugged mountainous backdrop; the seaside panorama to the south, to Kalk Bay and beyond, with the ever-active railway line on the right and the vision of spouting whales off-shore on the left; the stunning drama of the drive to Cape Point where two great oceans meet. It all made a vivid impression on me during my first visit to Muizenberg, and indeed my first visit to South Africa, for the meeting of the AIMS Council and the inauguration of AIMS in September 2003.

Anthony Pearson at Cape Point April 2004

Neil Turok's vision was already in place: that of creating a College-style centre of intense teaching, learning and research for gifted graduate students from all over Africa, to enable them to cross that difficult bridge from undergraduate study to independent creative thinking in the mathematical sciences. The first course had just begun, and the enthusiasm and outstanding motivation of the first cohort of students was plain for all to see. AIMS was to provide a beacon of hope radiating across Africa to aspiring graduate students eager to continue their studies to a level at which they might be able to engage in independent research for the benefit of their country and their continent.

I was honoured to serve on the Council of AIMS for 10 years, representing Cambridge University, which, together with Oxford and the Université de Paris-Sud, subscribed to a *Memorandum of Understanding* with AIMS in August 2003. Cambridge has been generous in support of AIMS through the teaching that it has provided; and I was particularly glad that Trinity College, at my instigation, provided a generous grant during the early years towards purchase of the properties on Melrose Road in Muizenberg (now housing the AIMS Research Centre) opposite the main AIMS building.

On a more personal note, I discovered a long-lost branch of my own family on my maternal grandmother's side, when I first visited Cape Town in 2003. My Grandmother Fleming, who died in 1976 at the age of 97, had always talked about her elder sister Jessie, who in 1891 had married a Scot, one John Todd Morrison. The couple emigrated to South Africa, where, I was always told, John Morrison became Professor of Physics at Stellenbosch University. Jessie had three sons, but she died in 1903 aged just 34, a tragedy that left its mark on

Reunion with my South African cousins
Cape Town, April 2004

her sister, my grandmother, even 70 years later. My visits to Muizenberg enabled me, with the kind help of Fritz Hahn, to track down the descendants of John and Jessie Morrison: four grandchildren, my second cousins, still living in Cape Town, with an extended network of their own children and grandchildren. We had a wonderful family reunion in April 2004. John Todd Morrison was indeed the first Professor of Physics at Stellenbosch, from about 1897. It is related that he advised Captain Scott, who stopped at Cape Town on his way to Antarctica, on the use of the latest magnetic compass; there is in consequence a Mount Morrison in Antarctica named after him!

Botswana 2006

After my visit to AIMS in January 2006, I spent a few days in Gabarone, Botswana, at the invitation of Edward Lungu, who had taken his PhD in Bristol in the late 1970s under my supervision. Edward was from Zambia, but had settled in Botswana as Mathematics Professor at the Botswana International University of Science and Technology (BIUST). He had also become actively involved in AIMS. I gave a seminar in Edward's Department, and had useful discussions with him and his colleagues.

IMS-Singapore

In March 1998, the Newton Institute had been privileged to receive a visit from Dr Tony Tan, then Deputy Prime Minister of Singapore, who invited the Institute to collaborate with the National University of Singapore in organising a millennial conference on *Fundamental Science: Mathematics and Theoretical Physics*. I was happy to be involved in this exciting venture, which came to fruition exactly two years later.

At the same time, Dr Tan told me about the planned foundation of a new Institute for Mathematical Sciences (IMS) under the aegis of the National University of Singapore, and modelled to some extent on the Newton Institute. In the event, I became a founding member of the Scientific Advisory Board (SAB) of IMS, which held its first meeting in December 2002 under the chairmanship of Roger Howe (Yale University) and hosted by the founding Director of IMS, Louis Chen. I served on the SAB till 2008; it met each year to discuss strategy and approve future programmes of the Institute. On one of these visits, I was honoured to be invited to tea with Lee Hsien Loong, Prime Minister of Singapore, in the Istana, the official seat of Government. Lee Hsien Loong had read mathematics at Trinity College, Cambridge, in the early 1970s, when I had supervised him for a term or two in Applied Mathematics. I recall a particular occasion during the miners' strike of January and February 1972, when there was a power cut and all the lights went out during an evening supervision; we continued by candlelight! Hsien Loong came top in Part II of the Mathematical Tripos in June 1973 (an achievement formerly distinguished by the title 'Senior Wrangler'). An ability to wrangle would serve him well in his future career as a politician! Hsien Loong was given the freedom of the City of London in 2014 at a splendid gathering at the Mansion House, the official residence of the Lord Mayor of London, that Linty and I were privileged to attend.

Founding members of Scientific Advisory Board of IMS (Singapore); (clockwise) Hans Fölmer, HKM, Louis CHEN (IMS Director), Roger Howe (SAB Chair), LUI Pau Chuan), CHONG Chi Tat (Management Board Chair), Avner Friedman, David Siegmund, LEUNG Ka Hin (IMS Deputy Director)

In 2006, following the meeting of the SAB, I took the bus up through Malaysia to Kuala Lumpur, where I met representatives of the local mechanics community in an attempt to persuade Malaysia to apply for membership of IUTAM. I note that Singapore became a member in 2021, but not yet Malaysia. My persuasive powers were clearly limited.

In April 2009, with the help and support of Emily Shuckburgh, I chaired a Spring School at IMS on the fluid dynamics of Typhoons, Tsunamis, Monsoon flooding and Atmospheric pollution, all matters of acute concern in south-east Asia. The School, aimed at Graduate Students from this region, was advertised by three posters in Chinese and English versions, designed by Andy Burbanks and printed by World Scientific, who also published the Proceedings in 2011.

The lectures were supplemented by afternoon project activity for the students working in groups of five or six. The photo here shows a group of students modelling the spread of a pollutant in a confined environment. The students were uniformly enthusiastic and engaged; and the weather provided fluid dynamical stimulus in the form of short heavy downpours most afternoons!

I supervise an afternoon project on spread of a pollutant

Emily Shuckburgh with a group of students at the Spring School

Midwest tour 2003

Hassan Aref invited me to undertake the 'Midwest tour', a rather hectic scramble giving seminars at nine Midwest universities -- the

University of Michigan at Ann Arbor, Michigan State University, Northwestern (in Chicago), University of Wisconsin, University of Minnesota, Notre Dame University (Indiana), Illinois Institute of Technology (IIT), University of Illinois Urbana-Champaign, and Purdue University (Indiana), -- over a two-week period; in retrospect it seems like a well-planned process of extended torture, involving several hours engaging with faculty members in each destination followed by my seminar and further discussion and (fortunately bibulous) evening entertainment. Each University chose one of the three seminar topics that I had been asked to offer, and fortunately, these were in the event evenly distributed.

In the course of all this, I was glad to make renewed acquaintance with Francis Bretherton, who had been a brilliant Research Fellow of Trinity in the early 60s, then a Fellow of King's. I had not met him since he left Cambridge in 1969 to take up a position at Johns Hopkins University, Baltimore. He was one of the first applied mathematicians to work on atmospheric dynamics, an emerging branch of fluid dynamics in the 1960s. What I most remember about Francis was that in discussions with him he would simply raise his voice until any countervailing opinions were effectively suppressed; it became a shouting match that he would always win. In USA, he became Director of the National Center for Atmospheric Research (NCAR), in Boulder, Colorado, where this talent may have served him well, before moving to the University of Wisconsin in 1980.

I had valuable conversations in the course of the tour with many other well-known personalities in the world of fluid mechanics including Ahmed Naghib, Julio Ottino, Fabian Waleffe, Tom Lundgren (1931--2021) and Dan Joseph (1929 -- 2011); I knew of their work, and was particularly glad to meet them personally. I kept a diary during the tour, and noted that there was much talk about Iraq, which had just been subjected to the the 'shock and awe' invasion of

George W. Bush. My American colleagues regarded him as a 'cowboy', and the question *"Where are these weapons of mass destruction (WMDs)"* was voiced in every conversation. Of course none were ever discovered. [Incidentally, the so-called 'dodgy dossier', which Tony Blair relied on in his decision to support the invasion of Iraq, was revealed by Glen Rangwala, Fellow of Trinity College, Cambridge, to have been seriously plagiarised from an article by Ibrahim al-Marashi, a postgraduate student at the Monterey Institute of International Studies, and to be irrelevant as this article concerned the situation in Iraq as it had been in 1991; see al-Marashi's captivating 2022 TedX talk *"The Dodgy Dossier, the Iraq War, and Me"*, on YouTube].

On a lighter note, my diary records that, at one stage in the tour, I found a launderette with the enticing signboard 'Ye olde washhouse'. There I washed my clothes, and accidentally left a few dollar bills in a pocket; I recorded that they came out nice and clean -- and this is how I discovered how to launder money!

In the late summer of 2003, Linty and I had a three-week adventure by car in France and Spain, first visiting our old friends Sandy Robertson and Francesca Bray at their hideaway in Mieres, Catalonia; then at Toulouse for the European Fluid Mechanics (EFM) Conference where I was awarded the first EFM Prize with its flattering citation; then to visit Linty's niece Jeanie who was living in a yurt near Laboule, in the high Cévennes with her two children and her latest paramour (another Keith!); then to Nice for the annual meeting of the French Society for Fluid Mechanics; and

Yurt life near Laboule, in the high Cévennes

finally for a few days holiday in Wimereux on the coast of Normandy on our way home to Cambridge. This was the summer of the 'canicule' ('cannie cool' might be the translation in Scotland!), and it sure was very hot, particularly in Toulouse.

Trinidad and Tobago

In January 2004, I went, accompanied by Linty, to Trinidad for the Caribbean Congress of Fluid Mechanics (CCOFD) at the invitation of Harold Ramkissoon, Professor at the University of the West Indies. We were delighted to run into Howard Stone at this same Congress. I had known Howard since his postdoc year in DAMTP in 1988/9. He was destined to win the first Batchelor Prize at ICTAM 2008 in Adelaide. Also here was Ibrahim Eltayeb, whom I had known since my visit to Khartoum in 1974. Ibrahim had collaborated for years with David Loper at Florida State University, so we had much common ground in geophysical fluid dynamics.

Following CCOFD, Linty and I escaped to Tobago, where we enjoyed a week of swimming, relaxing and reading at the Blue Waters Inn, in an ideal beachside location near Speyside (I wonder where that name came from): palm trees and banyans, bouganvillea in bloom, humming birds, sandpipers, fireflies, delicious food and pina-colada; what more need be said?

ICTP Trieste, 40th Anniversary

In October 2004, Katepalli Sreenivasan (Sreeni) invited me to lecture at a two-day Symposium to mark the 40th Anniversary of the International Center for Theoretical Physics (ICTP) in Trieste, which had been founded by Abdus Salam in 1964. The final lecture at the meeting, following a lecture by Cambridge economist Partha Dasgupta, was given by John Nash (of 'Nash equilibrium' and 'a beautiful mind' fame). His lecture with the obscurely uninteresting title "*An interesting equation*" turned out to be totally incomprehensible. He had prepared hand-written transparencies packed with so much detail that they defied comprehension. During the mandatory applause at the end of the lecture, a throng of female students flooded down from the back of the lecture theatre, quickly swamping Nash and demanding his autograph. The custodians had to swiftly escort him to safety. What a terrible tragedy it was when Nash and his wife were killed in a traffic accident on return from the award to him of the Abel Prize in 2015!

My 70th Birthday

Mike Proctor planned a two-day meeting 21/22 April, 2005, in DAMTP, with the title "*Turbulence, Twist and Treacle*" to mark my 70th birthday, *twist* relating to my discovery of *helicity*, and *treacle* relating to my work on slow viscous flow, particularly my discovery of corner eddies that came to be known as '*Moffatt eddies*'. Julian Hunt, René Moreau, Andrew Soward, Jens Eggers, Juri Toomre, Tim Pedley, John Hinch, Volodya Vladimirov, Renzo Ricca, Konrad Bajer, Yutaka Shimomura and Uriel Frisch had all agreed to lecture, as well as Mike himself. On 17 April, Mike's wife Julia phoned me to say that Mike was in intensive care in Addenbrooke's Hospital, with an allergic reaction (Stevens-Johnson syndrome) to an arthritic drug that he had been prescribed. Crisis situation! I e-mailed the authorities in Trinity, where Mike was at the time Dean of College, and in DAMTP where he was scheduled to lecture in the Easter Term, to inform them of the situation. Uriel Frisch agreed to take over as Chair of the meeting, due to start in 4 days time. I visited Mike every day that week in Addenbrooke's; it was a life-threatening situation, but fortunately his condition improved a little as time went by. The situation was dire but he faced the long period of slow recovery with great courage; it would take years for him to recover fully. Even so, he was elected Vice-Master of Trinity College in 2006, a position that he held with distinction.

My birthday meeting with about 50 participants, a wonderful array of former students and international colleagues, went ahead as planned, and we had a good dinner in Trinity on the Thursday evening. My good friend and colleague, Nigel Weiss, proposed my health, and I responded with a sonnet *Threescore Years and Ten* that I composed for the occasion (easier than a speech! --- see Chapter 10).

Université de Tous les Savoirs, Paris, 2005

I was invited that year to give a lecture on *La Mécanique des Fluides* at the Université de Tous Les Savoirs (UTLS) in Paris, actually on Saturday 18th June. This was a flattering invitation, and I had to accept. I was very tied up in Cambridge on the previous day, but booked seats for Linty and me on the last Eurostar of the Friday, which in those days stopped at Ashford (in Kent). We decided to drive to Ashford and take the train from there, and we set off in good time from Cambridge. I had prepared my lecture on Powerpoint, the first time I had used this, rather than transparencies, in an overseas lecture. But when we were well on our way down the M11, I realised I had forgotten the power cable for the laptop, which was going to be essential. I decided to turn at the Haverhill exit, and return to Cambridge for the cable. Now the timing became critical for us to get to Ashford in time for the train. I have never driven so fast either before or since, 100 mph most of the way. Amazingly, we were not caught in any speed trap. Linty was in more of a panic that I was. We arrived at Ashford, parked the car forgetting to close the windows in our rush, and were waved quickly through passport control. The train was in the station, and we leapt onboard through a carriage door that closed silently behind us. We sank breathlessly into two vacant seats, which miraculously turned out to be the very two that we had reserved. Champagne was served as the train took off! Incredible relief!

We stayed the night at our favourite Paris Hotel, l'Hotel des Grandes Ecoles, in the Rue du Cardinal Lemoine. Marie Farge came to see us on the Saturday morning, and to help me make final touches to my Powerpoint (one reason the power cable was essential) and to correct the French in which I had prepared it. And so finally, I was able to give the lecture that afternoon. Phew! Remarkably, that lecture, aimed at the scientifically curious Parisian public, is available online at < https://www.canal-u.tv/ chaines/utls/la-physique-fondamentale/la-mecanique-des-fluides >.

KITP Santa Barbara 2008

Linty and I spent the month of May 2008 at KITP Santa Barbara, during a programme on Dynamo Theory coordinated by Chris Jones, Daniel Lathrop, Steve Tobias and Ellen Zweibel. On one day during our stay, we made a boat trip from Ventura to the Channel Islands National Park with Krzysztof Mizerski, Sasha Ruzmaikin and Joan Feynman. Sasha had translated my 1978 Monograph into Russian, and claimed that he improved it in the process! He emigrated from Russia very soon after Perestroika in 1990, and married Joan in California some years later. Krzysztof is a keen sailor, and took me sailing on another day from the Santa Barbara Harbour. He would have sailed all night if I had not got nervous and persuaded him to return to port as the sun set!

Statue of James Clerk Maxwell, 2008

I had a less arduous trip up to Edinburgh for the unveiling of the statue of James Clerk Maxwell. This statue had been commissioned by Michael Atiyah with the sculptor Sandy Stoddart, following a vigorous fund-raising campaign. It stands at the east end of George Street, and serves to commemorate the greatest Scottish scientist of all time. Richard Feynman said in glowing terms that *"from a long view of the history of mankind, seen from, say, ten thousand years from now, there can be little doubt that the most significant event of the 19th century will be judged as Maxwell's discovery of the laws of electrodynamics"*. And Einstein said that *"the work of James Clerk Maxwell changed the world forever"*. In 2006, at a meeting in Paisley to celebrate the 'Year of Maxwell' in Scotland, I composed a poem *"The Genius of Glenlair"*. The final lines were read by Alex Fergusson, MSP, at the unveiling of the statue. Here are these lines (converted to English dialect):

Linty with (from left) Krzysztof Mizerski, Joan Feynman and Sasha Ruzmaikin, at Potato Harbor Vista Point, in the Channel Islands National Park, California

Blithe son o' Gallovidian hills, O' birk-clad slopes and tumbling rills,
Who rose through intellect sublime, To comprehend both space and time;
Great Scot! whose words in prose and rhyme, Inspire us yet o'er vales of time,
In this thine eponymial year Thy soaring spirit we revere!

DAMTP Jubilee, 2009

As the 'naughties' decade came to a close, we celebrated the 50th anniversary of the foundation of DAMTP, with a dinner in King's College attended by the Vice Chancellor Alison Richard and about 250 guests, including two of the great benefactors of the Department, Dill Faulkes and Peter Gershon. Speeches were given by the Head of Department, Peter Haynes, and by John Polkinghorne who had been Professor of Mathematical Physics from 1966-1979, before resigning his Chair to take up clerical orders. I also gave a speech in the course of which I paid tribute to James Lighthill and George Batchelor in the following terms:

"I look back on the 1970s as the decade of Lighthill in DAMTP. His was a flamboyant presence, and a great inspiration. Who can forget his seminar on the flight of the chalcid wasp Encarsia Formosa, which he mimicked with such enthusiasm that he appeared to float above the podium? ... or his account of his swim around the volcanic island of Stromboli while it was erupting? His great passion was swimming round islands, and he carried this to the ultimate extreme with his nine-hour swim round Sark in the Channel Islands at the age of 74; he died while still in the water through failure of the mitral valve in his heart, which had a weakness of which he had been unaware. This event had all the flavour of a Greek tragedy: Lighthill was our Icarus who had flown too close to the sun!

"I have spoken elsewhere of certain parallels between Batchelor and Lighthill on the one hand, and their 19th century counterparts, Stokes and Kelvin, on the other. In character, Batchelor and Lighthill could not have been more different; yet they shared a passionate interest in fluid mechanics that transcended their temperamental divergences, and ensured a decade of great achievement for their research groups here throughout

the '70s. We were truly privileged to have such intellectual giants of the subject in our midst!"

The Vice Chancellor made a concluding speech, that brought these Jubilee celebrations to a close.

CHAPTER 8

The 10s: Topological Dynamics

Bozeman, Montana 2010

A meeting was held in 2010 in Bozeman, Montana, to celebrate the 75th birthday of Steve Childress, who has made highly original contributions to fluid mechanics and dynamo theory over many decades. Some years previously, I shared social time with Steve at a meeting in Budapest, when we swam in the magnificent thermal baths at the Hotel Gellèrt. Steve's 1995 book "Stretch, twist, fold: the fast dynamo" arose from collaboration with my former student Andrew Gilbert that began during the Newton Institute programme on Dynamo Theory in 1992, and is still a leading text on the subject. From Bozeman, the participants enjoyed a visit to the Yellowstone National Park, and viewed its perpetually smoking geysers.

Steve Childress at his 75th birthday meeting, Bozeman, Montana, 2010

Warsaw 2011

The 13th European Turbulence Conference (ETC13) was held in Warsaw in September 2011. My former student Konrad Bajer was one of the organisers, and he also arranged a follow-up Symposium on *"Turbulence - the Historical Perspective"*. Konrad himself gave a beautiful introduction to this Symposium, which was recorded, as were all the later contributions which are available on YouTube.

On the Sunday following these meetings, Konrad and his wife Małgorzata invited me and Linty to spend the day with them at their dacha in the forest region just outside Warsaw. We had enjoyed a walk and a convivial lunch, when Linty's mobile phone rang, a very unusual event, as she used it only in emergency. Hester was on the line to say that Pierre Comte, Penelope's partner of 20 years, had committed suicide in Poitiers and Penelope was distraught. Linty immediately phoned Penelope to give her what comfort we could from that remote location. She had returned from La Rochelle that morning and found that Pierre had shot himself at their home in Poitiers; police were everywhere and she was temporarily barred from the house, but neighbours came to her rescue. Konrad had known Pierre for many years, and was as shocked as we were by this news. His vodka helped us to cope with this trauma.

Lunch with Konrad and Małgorzata at their dacha, September 2011; Linty took this photo just before we received the news of Pierre's death

Pierre had been due to participate at the Turbulence Colloquium held at the *Centre International de Rencontres Mathématiques* (CIRM) in Marseille just one week later. This meeting was organised by Marie Farge and Kai Schneider to mark the 50th anniversary of the seminal

1961 *Colloque International sur la Mécanique de la Turbulence* that I had attended, so my presence at this anniversary was essential. Ed Spiegel (1931--2020), the only other surviver of the 1961 meeting, was also present; he was himself still traumatised by the loss of his wife Barbara, on whom he had been very dependent until her death of a brain tumour earlier that year. I helped edit the Proceedings of the 2011 meeting, published two years later in the online *Journal of Turbulence*.

AIMS Senegal 2011

AIMS Senegal was established in 2011, the second AIMS after South Africa, and I arranged a visit at the earliest opportunity in January 2012. I flew to Dakar and was met by a taxi that took me the 45-mile drive to Mbour down the coast where AIMS is located. This was a two-hour drive on a pot-holed road in very heavy traffic with much hooting of horns on the way -- a great relief to arrive finally at the destination, in a beautiful forested location by the sea, the *Reserve Ecologique experimentale de Mbour*, where huge Beobab trees flourished.

AIMS Senegal in the *Reserve Ecologique experimentale de Mbour*.

I lectured there on mechanical toys, Euler's disc, the rising egg, and the rattleback, which I had with me for demonstration purposes; also a demonstration with a rolling magnetised ball that illustrates conservation of energy and momentum in a surprisingly dramatic way.

The return journey to Dakar was as hair-raising as the journey to Mbour, but now the taxi-driver simply left the road when faced with any hold-up and made a detour across the rough sandy scrub to circumvent it, so that we arrived in good time at the airport for my onward flight to Cape Town. But here, I ran into a different problem, for it appeared that I needed a yellow-fever vaccination certificate before boarding the flight. I had not known of this requirement previously. Anyway, the emigration officers kindly said they could have me vaccinated immediately, and they led me to a dingy room in the airport, where an even dingier 'medical officer' sat behind a desk. I was very apprehensive about being vaccinated in such a place, but he merely asked me for $10 and in return gave me an authentic-looking certificate which enabled me to board the plane with great relief and without further problem.

Topological dynamics 2012

Back in Cambridge, the programme *"Topological dynamics in the physical and biological sciences"* ran at the Newton Institute from July to December 2012. I organised this with the crucial support of Konrad Bajer, Tom Kephart, Yoshifumi Kimura and Andrzej Stasiak. Tom had worked (with Roman Buniy) on the energy spectrum of tight knots and links in its application to 'glueballs' in fundamerntal particle physics, while Andrzej had pioneered the application of knot theory in biological systems. I was working with Konrad in follow-up to the earlier meeting on *Turbulence* in Warsaw, and it was now that my still-ongoing collaboration with Yoshi Kimura on the interaction of vortex tubes got underway.

The six-month programme had a temporary hiatus for me in that in October that year I underwent a minor prostate operation, a 'HoLEP procedure' that was very successful for me personally; but it did put me out of action for a couple of weeks of the programme.

Konrad invited Andrzej Hercynski to spend a few days at the INI during the programme, and I was very happy to welcome this son of Ryczard Hercynski, who had spent a year's sabbatical in DAMTP away back in 1960, when I had first got to know him. Andrzej's elder brother Jannick had been one of my tutorial pupils at Trinity around 1970, so my link with the Hercynski family was strong. It remains so now through my close friendship with Andrzej, who ran a subsequent programme at the Newton Institute, *Growth form and self-organisation*, in 2017, and has been a regular visitor to Cambridge since then.

NAS 2013

My granddaughter Chloe was studying a degree course at the Berklee College of Music in Boston, and she was able to accompany me in Washington DC at the annual meeting of the National Academy of Sciences (NAS), to which I had been elected in 2008. I was very glad to meet there with old friends Andy and Jenny Acrivos, with whom we sat at the celebration dinner. Barack Obama gave an inspirational address at the meeting, and the Scottish violinist Nicola Benedetti. gave an equally inspirational performance at an evening concert.

Chloe at 21, in June 2013

In November that year, I was happy to welcome John Nott, who was visiting Trinity with his Slovenian wife Miloška and their granddaughter Saffron. This brought back conflicting memories from my early days in Trinity, 1957/59 when I had an anguished affair with Miloška, anguished because, all being fair in love and war, John swept Miloška off her feet and married her in 1959, thus abruptly bringing an end to our relationship. John graduated that year and went into finance and politics, becoming an MP in 1966. He attained fame as Defence Secretary under Margaret Thatcher during the Falklands War, about which he has written in his 2002 autobiography *"Here today, Gone tomorrow"*, to which Miloška also contributed (see also Chapter 9).

John Nott with Miloska and granddaughter Saffron in Nevile's Court, November 2013

AIMS Ghana and AIMS Cameroon 2014

AIMS Ghana was established in 2012 and AIMS Cameroon in 2013. The outstanding progress in the establishment of these new AIMS Centres was due to the infectious drive, commitment, and enthusiasm of Thierry Zomahoun, Executive Director of AIMS-NEI, and his support team.

I visited both Institutes in February 2014. I flew by Turkish airlines via Istanbul to Accra, and from there by car to AIMS, which was then located in Biriwa, near the Cape Coast and with a wonderful view over the Atlantic Ocean. This was only temporary, and it has since moved to central Accra.

View over the Atlantic from AIMS at Birawa, February 2014

I visit Cape Coast Castle accompanied by two members of the Staff of AIMS Ghana

I stayed only three days at AIMS Ghana, but was able in that time to visit Cape Coast Castle, the grim point of departure of slaves shipped to the Carribean. The underground dungeon was "a space of terror, death, and darkness", and it still carries a fearful atmosphere of that heinous crime against humanity, committed over a very extended period, truly an awful place that has to be seen to be believed. And it is sobering to consider that we in Western Europe are still the largely unwitting and unworthy beneficiaries of this abhorrent history.

From Ghana, with a 5-hour stop in Abidjan in the Côte d'Ivoire, I flew on to Yaounde, Cameroon. This was for the official launch of AIMS Cameroon, which itself is located at Limbe on the coast. The AIMS Board was received by the Prime Minister of Cameroon, Philémon Yang, who expressed his enthusiasm for mathematics in giving each of us a parting gift of a royal elephant head made from kola nut tree, representing "personality, wisdom and magnificence". This was large and heavy, and, for all its magnificence, created quite a problem on the flight home with the change of plane at Istanbul!

I get a warm welcome in Yaounde | The black sands at Bobende, near Limbe | Cameroon, inland from Bobende

In Limbe, I gave a 3-hour 'interactive' lecture on various dynamical systems, which the students seemed to find interesting; at least they did interact, asking many challenging questions. I spent an enjoyable day with Paul Wiegmann on the black volcanic sands at Bobende, where we enjoyed a dip in the surf, then explored inland on the very rough volcanic terrain.

Back in Cambridge, plans were afoot in Trinity for a major renovation of New Court, and it was evident that I would soon have to clear my room there. I decided to send my complete set of volumes of JFM to AIMS Ghana, in the hope that this might stimulate research activity there in fluid dynamics; and I sent many books also to AIMS Cameroon, which was also starting up a small library.

My beloved set of volumes of *JFM* shipped to AIMS Ghana

Konrad Bajer, deceased 2014

Konrad Bajer had visited Cambridge in November 2013, and was helping me to plan a new edition of my 1978 Monograph *"Magnetic field generation in electrically conducting fluids"*, which was long overdue. I was in regular e-mail contact with him in the following months, and

I was increasingly concerned about a persistent cough that he had which just refused to go away. He died in Warsaw on 29th August 2014 at the age of 57 from a rare form of cancer, diagnosed only one week before his death. He was in fact working on his laptop from his hospital bed until the very morning of his death, when he wrote and e-mailed to Cambridge a glowing testimonial for one of his students! Another sad sad day, hard to endure. Milton's words come to mind:

With Konrad in Cambridge, 2007

Bitter constraint, and sad occasion dear,
Compels me to disturb your season due;
For Lycidas is dead, dead ere his prime,
Young Lycidas, and hath not left his peer.
Who would not sing for Lycidas? He knew
Himself to sing, and build the lofty rhyme.

I attended Konrad's funeral in Warsaw on 5th September 2014, and was able to pay tribute at his graveside to the great friendship that we had enjoyed.

Giffen Goods

My son Peter is Professor of Econometrics at the University of East Anglia. Occasionally he consults me about mathematical problems that he runs into. One of these concerned Giffen Goods, which have the unusual property that the demand for such goods increases as their price increases, other prices being held fixed, in contrast to the 'normal' response. My favourite example is that of a hypothetical Scot who spends all his disposable income on oatmeal and malt whisky, the latter being a relatively expensive luxury. If the price of oatmeal increases due to persistent rain and a poor harvest, our Scot can then afford less whisky, but, needing to maintain his intake of calories, must compensate by buying more oatmeal, which is in these circumstances a Giffen good. We published our paper *Giffen goods and their reflexion property* in the *Manchester School* 2014.

Pete at work on Giffen Goods

80th birthday

The year 2015 inexorably led to my 80th birthday in April. It is the good custom in Trinity College that the Master proposes a toast to the new octogenarian, who then has to respond with a speech covering highlights of his or her childhood background and subsequent career. My family were invited for this occasion and Greg Winter, as Master, proposed the toast. I gave my speech which has been printed in the 2015 Annual Record of the College. I reproduce here only the frivolous concluding paragraph, and the ballad to which it refers:

"*Master, I would like to conclude by reciting to you, with your consent, or indeed without it, a version of an old Scots ballad, "The Twa Corbies", which, being translated, means "The Two Crows"; some of you may be familiar with this, or with an inflationary English version "The Three Ravens". This version that I shall recite was discovered on a scrap of parchment during an exploration of the Whewell's Court cellars that I conducted with Dr Seal in October 1967, with a view to possible renovation;*

the parchment has been carbon-dated with remarkable accuracy to the 1st of April 1865, when Whewell's Court was under construction. This ballad has sombre undertones; I doubt that I can read it with the appropriate level of solemnity. But it also has historic interest in that it provides clear evidence that the practice of the 'buzz' in the Combination Room was already well established by 1865."

The Twa Corbies

As I ga'ed oot on the Kings' Parade
I spied twa dons wi' thir gowns displayed;
The ane unto the tither did say
Whar sall we gang an' dine the day?

In across yon cobbled court,
I wot they serve a goodly port,
From dungeons deep beneath the stair,
And naebody kens that it lies there.

Thon Fellow wi' the shredded gown —
We'll hem him in and wear him down:
He'll drain the flask — he always does,
For that's the way he gets a buzz[1].

We'll sit on as the candles wane
An' conspire to pick his brain;
One good idea is all we need
An' wir next joint paper is guaranteed.

Mony a one for him maks mane,
But naebody kens whar his mind has gane;
In the Great Court he'll rend his hair,
An' the fount shall spout for evermair.

Research continued

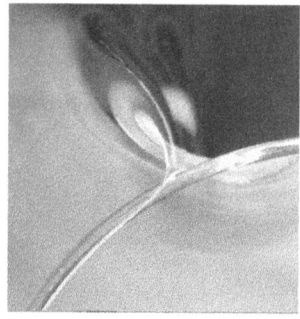

Visualisation of the twist singularity on a boundary wire

I was very happy to have a paper *"Flashpoint"*, joint with Donald Lynden-Bell, published in *Monthly Notices of the Royal Astronomical Society* in the year 2015 in which we both achieved octogenarian status. During our work together on that paper, we were able to reminisce about our adventures at the 1959 Les Houches summer school, so many years earlier. It was also about now that my collaboration with Ray Goldstein and Adriana Pesci began. Renzo and I had been puzzling over the behaviour of a soap film in the form of a Möbius strip. You can create this by twisting a circular wire and folding it back on itself, as you might do in 'doubling' an elastic band. The wire then forms the boundary of a Möbius strip, and a 'one-sided' soap film can be easily formed by dipping this in soap solution. If now the wire is carefully unfolded and untwisted back to circular form, then at a critical instant, the one-sided soap-film jumps to a conventional two-sided, disc-like form, an interesting topological transition. The question then arises as to how exactly this jump occurs. Ray and Adriana were able to resolve this with high-speed photography, revealing the twist singularity that appears on the boundary wire at the moment of transition.

Venice 2016

The highlight of 2016 for me was actually the IUTAM Symposium on Helicity, Structures and Singularity in Fluid and Plasma Dynamics for which the moving spirit was my former student, Renzo Ricca, now Professor at the University of Milano-Bicocca. Renzo had originally planned this to coincide with my 80th birthday, but the planning slipped for a year to the week of my 81st birthday, 11-15 April 2016, when it was held in the magnificent Palazzo Franchetti in Venice, secured for the Symposium through the good offices of Bernhard Schrefler, who was at that time a member of the Bureau of IUTAM. I joked with Renzo that 81 (= 3^{2^2}) was a much more significant number than 80, being just the second number in the sequence $(n+1)^{n^n}$, (for n=1, 2, 3, ...); [the next number in this sequence is 4^{3^3}=18014398509481984]. Anyway, it was a brilliant Symposium. My only regret was that Konrad Bajer, who had been involved in the early planning, but who died in 2014, could be present only in spirit!

With Marie Farge on our hotel balcony, Venice, April 2016

Emmanuel Dormy whom I had known since my Blaise Pascal days in Paris, 2002, was at this Symposium with his partner Ludivine Oruba, and it was about now that Emmanuel and I began collaborating on that long-delayed revised edition of my 1978 monograph. In the event, the new material that we included was so extensive that David Tranah, Science Editor at CUP, recommended that it be treated as a new book; thus it was that it was finally published under our joint authorship in

With Renzo, in Venice, 2016

2019, and with the new title "*Self-Exciting Fluid Dynamos*". There were inevitably some loose ends that have kept us busy ever since.

The finite-time singularity problem

My collaboration with Yoshi Kimura was also beginning to bear fruit, and it was now that we launched our attack on the notorious 'finite-time singularity problem' for flow governed by the Navier-Stokes equations. The motivation for this study is very clear: on the one hand, the now classical Kolmogorov theory of turbulence requires that the rate of viscous dissipation of the energy of the flow should remain finite in the limit as the fluid viscosity tends to zero. This in turn requires that the field of vorticity (related to the velocity gradient) must exhibit some kind of singular behaviour in this limit. However, no explicit example of the finite-time development of singularities starting from smooth initial conditions of finite energy has yet been found, and most mathematicians believe that such behaviour simply cannot occur. So there is a contradiction here that strikes at the very heart of fluid dynamics, and particularly at the theory of turbulence, rightly described by Feynman as "the most important unsolved problem in classical physics".

Yoshi and I studied the behaviour of two vortex rings oriented to propagate towards each other on inclined planes. We believed for a while that this configuration could exhibit singular behaviour, but in this we were mistaken for reasons that are too subtle for me to attempt to explain here. This however is an area where excitement at the prospect of a breakthrough alternates with depression in seeing only a brick wall ahead.

Whittaker Colloquium 2017

With Sir Michael Atiyah at the Royal Society of Edinburgh, October 2017

Sir Michael Atiyah invited me to give the Whittaker Colloquium in Edinburgh in 2017, and this I was very happy to do. Edmund Whittaker (1873-1956) was a Fellow of Trinity College from 1896-1906 and held the Chair of Mathematics at Edinburgh from 1912 till 1946, when he was succeeded by Alexander C. Aitken. This Colloquium, which was video-recorded and is accessible on YouTube, gave me an opportunity to reflect on my own time as a student at Edinburgh University in the 1950s before an Edinburgh audience. My lecture on *Topological Fluid Dynamics* also allowed me to dwell on the seminal investigations of Kelvin, Tait and Maxwell in the 1860s and 1870s, which provided the basis for my own researches a hundred years later.

Atiyah's wife Lily died in 2018, and Michael arranged a Memorial meeting in the Playfair Library, off the Old Quad of Edinburgh University. He asked me to write a poem for the occasion. I composed a sonnet praising her own early achievements in mathematics under the supervision of Mary Cartwright. Michael himself

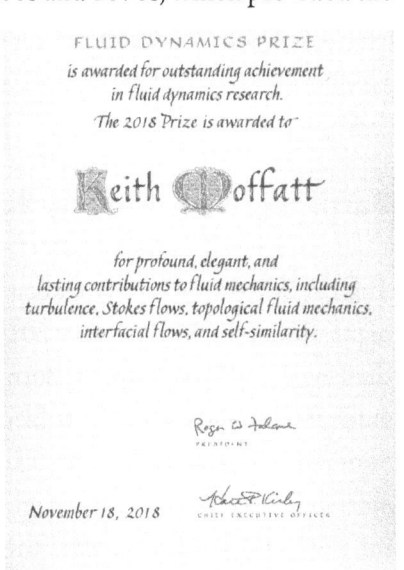

died just one year later; and so passed "probably the greatest British mathematician since Isaac Newton", as judged by Ian Stewart in his Guardian obituary.

APS Fluid Dynamics Prize and Otto Laporte Lecture 2018

In 2018, I was awarded the Fluid Dynamics Prize of the American Physical Society (APS); this prize was coupled with the Otto Laporte lecture, which I gave at the APS meeting in Atlanta in November that year. On the flight out, I discovered an error in the computations that Yoshi and I had carried out on the finite-time singularity problem, which was the main part of my lecture. Panic ensued! Fortunately I had three days in Atlanta before my lecture, so that, aided by the jet-lag that made me wake at three in the morning for the three days, I was able to correct this error and revise my presentation accordingly. [Fazle Hussain and his postdoc Jie Yau discovered another error later, which we acknowledged in a published Corrigendum; such are the hazards encountered in the rush to publish too precipitately!].

Burns Night 25 January 2019

The BA Society of Trinity College were active in organising a Burns night dinner on 25th January 2019, complete with a piper to pipe in the Haggis, and recitation of Burns' *Address to a Haggis* by one of our graduate students from Aberdeen, which she did with great effect. Someone suggested that I should deliver the Selkirk Grace "*Some hae meat an' canna eat; some hae nane an' want it; we hae meat an' we can eat, sae let the Lord be thankit*", this to replace our usual Grace "*Oculi *omnium in te sperant domine*, . . ." Such a profound change must have been agreed by the College Council, and so I found myself delivering the Selkirk Grace much to the astonishment of the gathered company, Fellows and students alike, that evening. It was after all the first time since the foundation of the college in 1546 that such a shocking substitution had been proposed, far less permitted. Surely a sign, if such sign were needed, that "the times they are a-changin' ". Nobody complained after the event; perhaps it was the whisky served with the haggis that provided sufficient mollification.

Oman 2019

One further event in 2019 deserves mention. My good friend Volodya Vladimirov had for some years held a position at the Sultan Qaboos University (SQU) in Oman. I had first met him in Novosibirsk, one of the coldest places on earth, and here he was now in Oman, one of the hottest. He certainly was a man of extremes! He was eager to establish research relations between SQU and Cambridge, and to this end he organised a three-day Symposium in Muscat that David Hughes (from Leeds), Mike Proctor (by now Provost of King's College, Cambridge) and I, were able to attend. Volodya was aided by Ibrahim Eltayeb, who had held a long-term position at SQU since emigrating from Sudan some 40 years earlier.

Volodya, searching for Frankincense in Oman

The heat was oppressive, but Volodya and his wife Natasha would take a walk early every morning in the relative cool before sunrise, and I joined them in this. Their home was air-conditioned, as were all lecture rooms at the University.

Bailebeg Speyside

Linty arriving at the cottage that we rented in Bailebeg

... and finding what looks like a stone-age rattleback by the nearby river Avon

By way of contrast, Linty and I spent a week later that year on holiday in Bailebeg, just south of Tomintoul and as near as we could get to the home of my Gaulrig ancestors. The view from the cottage was of Highland cattle curious to inspect us on our arrival. I managed to walk to Gaulrig, and was glad to see that a cottage had been restored for occupation by a solitary gamekeeper. I found many ruins, some of which might well have been ruins of the illicit stills that were the source of good Scotch whisky long before the 1820s, when the distillation process was legalised and the product taxed. We visited the Glenlivet distillery, where bottles are arranged in a fine helical array: helicity with a difference!

CHAPTER 9

The 20s: Pandemic Preoccupations

Timelessness and Eternity

During my first year at Trinity as an affiliated student (1957/8), there were two chaplains in the College, Simon Phipps (1921--2001), later Bishop of Lincoln, and Eric James (1925--2012), each being assigned half of the undergraduates of the College for their spiritual well-being and moral support, as and when required. Eric was 'my' chaplain who became a close friend and advisor long after he left Cambridge. He was for many years a contributor to the *Thought for the Day* programme on Radio 4 and actually devoted one of these to the 'abutilon miracle' that I related to him following the death of Fergus in 1987.

I recall an evening in Trinity when Eric invited a small group of students to his room for a discussion (over a glass of wine, to be sure) led by Peter Long (1925--2018), a Research Fellow of the College. Peter's field of research was Philosophy, and he chose as his theme that evening Wittgenstein's statement *"If by eternity we mean timelessness, then he lives eternally who lives in the present"*. This quotation has remained with me for a lifetime, and I thank Peter for that. It is of course a big IF; could I not equally say "If by chalk we mean cheese, then he who works at the blackboard may find writing on it a bit troublesome"? In offering this comparison, I lean on another quote attributed to Wittgenstein: *"Never stay up on the barren heights of cleverness, but come down into the green valleys of silliness"*. So be it! Ludvig Wittgenstein (1889--1951) was himself a Fellow of Trinity from 1939 until his death in Cambridge in 1951. He lived in a spartan room in Whewell's Court with minimal furnishings, and it is said that he seldom dined at High Table, because he found the conversation too trivial.

Anyway, I was so enthralled by the discussion with Peter Long that I sought his company, and I used to find him at the Coffee Pot in Green Street, where we would discuss matters of life and death over coffee and, yes, Gauloises. This was much more enjoyable than attending mathematics lectures at 9 in the morning!

The quote of Wittgenstein comes back to me now, as I approach my 90th birthday. Nine decades is my eternity, and yes, there is now a certain timelessness about it. I must take each day as it comes and indeed live in the present, that instant at which the future seamlessly fades into the past. In this spirit, and with the help of a dram or two of Single Malt, I summarise this most recent period of my life.

And yet, still digressing, as I contemplate eternity, the first verse of my poem "Genesis: Cosmological Echoes" comes to mind:

> *There was a time when Time itself stood still*
> *And Triune Space was formless, void and vast;*
> *No matter stirred, there was no eye to see*
> *Nor mind to comprehend the vacant past.*

In this, I was trying to convey an impression of nothingness, a state of 'timelessness' which could well be considered as an 'eternity'. Perhaps unwittingly I was in tune with Wittgenstein after all. I include in Chapter 10 the full version of this poem, which comes as near as may be to a statement of my own religious leanings.

The Pandemic

The Covid pandemic hit the UK on 20 March 2020, when the first lockdown was imposed. On that very day, I gave a zoom seminar *"Some Topological Aspects of Fluid Dynamics"* on GEOTOP-A, the Web-seminar series on *Applications of Geometry and Topology* that had been initiated by Renzo Ricca, De Witt Sumners, and others in August 2018.

For the next two years, we were to be limited to zoomed lectures, seminars and conferences, in the worldwide attempt to control the spread of the virus. Even the ICTAM that had been planned for Milan in August 2020 was first postponed for a year, and then reduced to a zoomed (one might almost say doomed) Congress.

We in Cambridge had planned to hold an IUTAM Symposium *"Fluid Mechanics in the Spirit of G.K.Batchelor"* to mark the centenary of Batchelor's birth in March 2020, but this too had to be postponed for a year, then converted to a zoomed Symposium. which was deliberately limited to nine invited lectures over three days. My good friend and photographer Howard Guest, with some input and encouragement from me, produced a special 'programme notebook' that was distributed to all registered participants, as a souvenir of this unusual event.

About now, my right eye began to trouble me. A mistiness appeared in the centre of my field of vision, and would not clear. My optician sent me to the eye clinic at Addenbrooke's Hospital (called the Emmeline clinic -- my mother's name -- although no-one there seems to know why), and retinal vein occlusion in the right eye was diagnosed. Over the next two years, I had numerous eye injections and laser treatment which gave very temporary relief, but it was finally decided that the condition was incurable, and I was prescribed daily drops for both eyes to prevent any worsening of the situation. So far, so good!

Barcelona 2024

Demonstration in la Rambla, Barcelona in support of Julian Assange, Jan 2024

Travel was very restricted during the pandemic, but the situation was beginning to ease by 2023. I was roped in as an assessor for ICREA, the Catalan Institution for Research and Advanced Studies, assessing applications for academic promotions at Catalan Universities. Much of the work was done online, but the final decisions were taken by committees that met in Barcelona in January 2024. I made the trip and much enjoyed this visit and interactions with a highly talented group of assessors drawn in from across Europe.

80th birthday gathering for Andrew Soward, here flanked by Emmanuel Dormy and myself in the front row; Steve Childress is on the extreme right.

In January 2024, I also attended a meeting in Newcastle to celebrate the 80th birthday of Andrew Soward, one of my former students who has had a spectacular career in Dynamo Theory, and as Editor of the journal *Geophysical and Astrophysical Fluid Dynamics* (*GAFD*).

Fingering instability

My research work continued intermittently during the pandemic. But, constrained as I was to work from home in 2020, I set up a simple experiment in my garage -- a demonstration that I had frequently used in lectures. This involves squeezing a drop of viscous fluid between two glass plates, a 'Hele-Shaw cell'. When the squeezing force is constant (as for example if the upper plate descends under its own weight), the drop becomes circular with a radius that increases as the one-eighth power of time. If the plates are then forcibly separated, a dramatic 'fingering' instability occurs, of a type first identified by Philip Saffman and G.I.Taylor in 1958. My photographer friend,

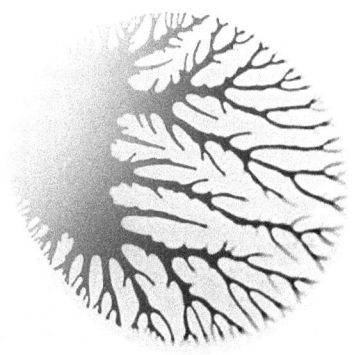

The fingering instability that develops in a drop of honey when the two glass confining plates are separated by a knife slipped in at the lower right-hand corner. photo © Howard Guest

Howard Guest, set up a greatly improved system in his home, with high-speed photography to follow the development of this instability. Herbert Huppert joined us in this investigation, which was published in *JFM* in 2021.

Frontiers in Dynamo Theory 2022

Shortly after the 2018 publication of our book "*Self-Exciting Fluid Dynamos*", my co-author Emmanuel Dormy was an organiser of the Newton Institute programme "*Frontiers in Dynamo Theory: from the Earth to the Stars*", which ran from September to December 2022. This gave us the opportunity to continue our research collaboration, which is still ongoing. A further programme "*Anti-diffusive dynamics: from sub-cellular to astrophysical scales*" ran from January to June 2024, and enabled me to continue to collaborate with Yoshi Kimura on aspects of vortex interaction, work that we presented at ICTAM 2024 in Daegu. I owe much to Emmanuel and Yoshi who continue to encourage and stimulate me to collaborate in continuing research investigations.

Miloška

Miloska in Cambridge, 1958

As I write today, 7 November 2024, I have just read the Times obituary of Sir John Nott, who died yesterday at the age of 92. As I mentioned in the previous chapter, John had come to visit me in 2013 in Trinity together with his wife Miloška and granddaughter Saffron. John's obituary refers to his time as a Trinity undergraduate in the following terms: "*In 1956 he went up to Trinity College, Cambridge, where he read economics and law. He already had a career in politics in mind. The main gain from his Cambridge time was marriage to Miloška, a striking Yugoslav ex-communist. They met at her engagement party to a Cambridge don and he [John] told her after five minutes that he was going to marry her. Four days later they married."* There is a grain of truth in this; but only a grain. It behoves me now to set the record straight in at least some respects, since I am the alleged "Cambridge don" in the above excerpt.

I arrived in Cambridge as an affiliated student in 1957, one year later than John, who was then three years older than me, having served six years in the army, three of them in Malaya with the Gurkha Rifles during the prolonged Communist insurgency. I first met Miloška at the home of Bill and Nina Wedderburn in Millington Road, where my sister Lindesay was lodging. Lindesay was employed at the University Library, having completed a degree in Classics at Edinburgh and a Diploma in Classical Archeology in Cambridge. Miloška was an '*au pair*' with the Wedderburn family, and studying English at the Studio Language School in Cambridge. She was exotic, being from Slovenia, and having somehow emerged from behind the Iron Curtain, to spend a year or so in Italy before coming on to Cambridge. I was smitten at first sight in January 1958, and we spent much time together over the following months.

John was meanwhile President of the Cambridge Union, and therefore a well-known personality in the student population. I took Miloška to a meeting of the University Explorers' and Travellers' Club (yes, there was such a Club) at which he gave a talk about his experiences in the Far East. I invited him to a party that I gave in the old Bevan Hostel in Green Street, long since demolished, where I was lodged during my first year in Cambridge. This was in May 1958 and I remember it well. I took over three rooms in the hostel and 40 friends showed up. Miloška helped me prepare canapés and punch, and of course we had lots of music, mainly of the country folk style, popular at that time. John did indeed spend most of the party trying to monopolise Miloška; but we were not then engaged. We did become formally engaged five months later in October 1958. Miloška came with me to spend Christmas at my home in Scotland, and charmed the family, to the point at which my mother Emmeline described her in her diary as "captivating and winsome, with a lovely smile and manner"; and plans were then well underway for us to marry in June 1959.

However, things began to go wrong in the New Year 1959. I was already in my first year of a PhD, and I was giving a maths supervision one evening in the Lent Term, when John burst into the room asking my permission to take Miloška out for the evening. I said that if she wished to go, I could not say no. This was perhaps the beginning of the end for me (and the end of the beginning for John). Miloška became more and more unsettled, and was even in March planning to spend a year in USA before marriage --- or so she said. I was depressed about the situation, as was plain when I went home alone for the Easter vacation. It was then that the bubble burst: I received a letter from Miloška saying that she had run into John in London, and they had got married. Just like that! Well, as I said, all's fair in love and war, but it took me some time to accept the situation. And it was Linty that finally rescued me a year later, when I realised that all was indeed for the best in the best of all possible worlds.

CHAPTER 10

Paradise enow

Wilderness were Paradise enow

> *A Book of Verses underneath the Bough,*
> *A Jug of Wine, a Loaf of Bread—and Thou*
> *Beside me singing in the Wilderness-*
> *Oh, Wilderness were Paradise enow!*

Thus spake Omar Khayyám as rendered by Edward Fitzgerald in 1859, and it stirs my heart to this day. And so with Dylan Thomas's *Do not go gentle into that good night,* and then *rage, rage against the dying of the light,* At the age of 39 he did not go gentle, but he went with poetry in his soul. So would I wish to go, with poetry in my soul. But I can rage too against the dying of the light, whether in Ukraine or in Gaza, with Shakespeare's *Woe to the hands that shed this costly blood.* Much poetry remains with me from schooldays at George Watson's College, and helps dream away the hours of sleeplessness; and not only *Wee sleekit cowrin tim'rous beastie!* In this *season of mists and mellow fruitfulness* that we currently enjoy, I can still thrill to that *wild west wind, thou breath of Autumn's being.* And who needs to travel when you can recall that *in Xanadu did Kubla Khan a stately pleasure dome decree,* or sense the *hushed Chorasmian waste* of Sohrab and Rustum. And it wasn't only in English lessons that we were tutored in poetry at Watson's. We had French too: *Heureux qui, comme Ulysse, a fait un beau voyage,* and Racine's *C'était pendant l'horreur d'une profonde nuit* with its horrifying climax:

> *Et moi, je lui tendais les mains pour l'embrasser.*
> *Mais je n'ai plus trouvé qu'un horrible mélange*
> *D'os et de chairs meurtris et traînés dans la fange,*
> *Des lambeaux pleins de sang et des membres affreux*
> *Que des chiens dévorants se disputaient entre eux.*

And Latin too, with Virgil's *Primus ibi ante omnis, magna comitante caterva, Laocoon ardens summa decurrit ab arce* leading to the immortal line *Timeo Danaos, et dona ferentes* (I fear the Greeks, especially when they are bringing gifts).

So, from the sublime to the ridiculous, here are some of my own attempts at rhyme and rhythm. First, note that, *If you're a bard, it's very hard/ To find a word that rhymes with rhythm,/ But in the end, you'll meet a friend,/ And maybe find it chatting with 'im./ You'll note this rhyme shows gender bias,/ And that's 'cos rhyming's meant to try us;/ But don't despair, nor droop, nor dither,/ Just find a friend and chatter with 'er./ Now that balance is restored,/ And oil on troubled waters poured,/ I'll dare to try my hand at rhyme,/ Please browse the rest if you have time.*

Thus did I introduce a booklet of poetry that my good friend Renzo Ricca had encouraged me to prepare for distribution at my 81st birthday Symposium in Venice, 2016. I reproduce here with some explanatory notes, some of the poems in that booklet.

Tam-day musings

On 7th August 2001, I received the following widely circulated e-mail from the late Hassan Aref, who had been President of ICTAM 2000 the previous year in Chicago:

Dear Fellow Mechanician,

Last year, as part of ICTAM 2000, Governor Ryan declared Monday, August 28, 2000, "Theoretical and Applied Mechanics Day" in the State of Illinois. We propose that henceforth the last Monday in August be designated "TAM Day". This year we encourage you to celebrate TAM Day on Monday, August 27. Please let us know how you plan to mark this occasion. Your creative ideas will be featured on our website **here.** *Happy TAM Day 2001!*

Department of Theoretical and Applied Mechanics, University of Illinois, Urbana-Champaign.

I was at this time approaching the end of my 5-year stint as Director of the Isaac Newton Institute, behind which we had recently installed a boules court as a recreational facility for the many visitors to the Institute (among whom at the time were Vladimir Zakharov and Martin Kruskal). I responded to Aref's directive with the following poem. Those with Scottish sensitivities will find the rhythm familiar, and will appreciate the level of plagiarism!

The boules court at the Newton Institute, 2001

A poem to celebrate TAM-day, 27th August 2001
After Tam O'Shanter, with apologies to Robert Burns

When lecture rooms are growing stale,
And thoughts begin to turn to ale;
When blackboards thick in chalk are smothered,
With scarce a new truth there discovered;
When loud the arguments have railed,
But longed-for inspiration failed;
Why, that's the time to take a break,
To turn to coffee, tea and cake;
Or sip the vineyard's produce cool,
And contemplate the game of boules.

This truth was haply brought to mind,
By letter sent and duly signed,
Conveying tidings of great joy
Hot from the heart of Illinois;
There it was formally declared
What ne'er had hitherto been dared:
Mechanics as a worldwide art
Should henceforth have a day apart,
A day when man may ruminate
Upon the subject's vibrant state;
A day when blackboard toil should cease,
And staff should have a bit of peace!
Straight from the land of Al Capone,
Instruction came by telephone;
It seemed there was no time to lose,
Such offers one cannot refuse!

And thus it was TAM-day was born,
An August Monday to adorn,
A day this year decreed by heaven
To fall on August twenty-seven;
So three times three, again times three,
A date on which TAM holds the key
To open Archimedes' door,
And celebrate on every shore;
What time of day? You well may ask;
The answer's plain: from dawn till dusk.

In Cambridge town we heard the call,
And rallied to the central hall
Of mighty Newton's Institute,
Irreverently called by some the Newt!
He who bestrode the pebbled shore
Of Ocean's ever-mobile floor;
Here a pebble, there a pebble,
But was the system integrable?
Said Newton "If I shed a tear
Upon this ocean wide and clear,
Will this affect the rain in Laos?"
And thus were sown the seeds of Chaos.

But to our tale: the day dawned bright,
The weather forecast had been right;
The warming sun on boules court shimmered,
The overarching crane fair glimmered;
That day a child might understand
An awesome drama was to hand.

Well practiced in the laws of motion,
And fortified by vintage potion,
The gifted savants slow foregathered
Hard by the shed where bikes are tethered;
From every land and clime they came,
Experts of legendary fame;
From Russia, It'ly and Japan,
And every country known to man;
From Poland, Greece and USA,
A clash of titans underway!

But here my Muse her wing maun cour,
This glorious game made such a stoor;
The boules of glittering steel were round,
And sped unerring on the ground;
They rolled, they arched, they spun, they clickit,
More action here than seen in cricket!

The game might well have run till dawn
There by the side of Newton's lawn.
The cochinet brent new frae France
Was kissed with steel, and touched perchance,
Balls tossed with Zakharovian skill;
Then Kruskal clad in T-shirt still,
He who could aim a ball and roll-it-on,
Invariant as any soliton,
With concentration ever keener
Entered upon that tense arena;
And cast one ball, a crafty throw
That scattered those of every foe.

The cochinet was split asunder,
Th'encircling crowd was mute with wonder.

Now, who this tale o' truth shall read,
Mechanics of whatever creed:
When overwork becomes a grind,
And problems tangle up the mind,
Fill well the cup and fill it full,
Take refuge in a game of boules!

James Clerk Maxwell, The Genius o' Glenlair

James Clerk Maxwell was a Fellow of Trinity College in the 1850s and again as an Honorary Fellow, in the 1870s. His portrait hangs in Hall. He is famous for having recognised that light is a form of electromagnetic radiation, and in this discovery he anticipated, and provided the basis for, Einstein's Special Theory of Relativity. The year 2006 was the 175th anniversary of his birth, and it was declared the "Year of Maxwell" in Scotland, the land of his birth. I wrote this poem to mark this anniversary. It is in the Scottish vernacular that would have been very familiar to Maxwell, certainly as a child in Galloway in the 1830s. Maxwell wrote poetry himself as a pastime; two of his poems are included in the 2016 anthology "Trinity Poets", edited by Adrian Poole and Angela Leighton, to which my poem is added as an Appendix -- a great honour to be included in that volume! [An annotated version appears on my personal website.]

When James Clerk Maxwell was a lad,
His questing mind fair deaved his Dad;
For "What's the go of it?" he'd speir,
An' pester on till a' was clear.

They ca'd him 'dafty' at the scule,
An' that, ye'd think, was awfie cruel!
He didna' mind, he was apart
Constructing ovals o' Descartes!

He played wi' colours blue an' green
An' red, enhanced by glorious sheen;
An' took the earliest colour photo,
As good as ony Blake or Giotto.

He analysed the rings o' Saturn ,
Resolving their striated pattern,
Predicting weel their composition
By calculus an' long division.

Redundant in the granite city
An' spurned by En'bro', mair's the pity,
He had tae traipse awa' doon Sooth,
Nae doot they thocht him gae uncouth!

He liked tae doodle lines o' force,
Wi' charge an' current as the source;
As much at hame wi' rho an' phi,
An' E an' B an' J forbye!

Through these he dreamt up waves o' licht,
An' workit on them day an' nicht;
His mind roamed far whaur ithers durn't,
An' hit upon displacement current.

Syne back tae Galloway he repaired,
He had tae go – he was the laird!
By day conferring wi' the ghillie,
An' lucubrating willy-nilly!

At last frae Cambridge cam' the call
Doon tae thon hallowed Senate Hall,
Where, tho' he held the dons in thrall,
They didna follow him at all!

Blithe son o' Gallovidian hills
O' birk-clad slopes an' tumbling rills,
Wha rose through intellect sublime,
Tae comprehend baith space an' time;

Great Scot! wha's words in prose an' rhyme,
Inspire us yet o'er vales o' time,
In this thine eponymial year
Thy soaring spirit we revere!

On the Planting of Newton's Apple Tree, a genuine descendent from the original tree at Woolsthorpe Manor, 19th November 2001, at the Newton Institute.

When Newton pondered 'neath the apple tree,
And hidden truths of Nature did discern,
Three universal laws he did decree
That those who seek might ever heed and learn;
There in the orchard did his mind take flight
O'er vistas wide, where only he could dare;
And to the planets having raised his sight,
Resolved their orbits through the inverse square.
This tree, transported here from leafy glade,
Of self-same strain that tasted Woolsthorpe's dew,
Well planted now through grace of silvered spade
May stir th'enquiring mind to conquests new,
Symbol of arts well-nurtured at the root,
That may through budding genius harvest fruit.

Linty wielded the silver spade

Dinner at Corpus

My brother-in-law, Kay Stiven, was an alumnus of Corpus Christi College, and had invited me as his guest at an Alumni Gathering of the College; here is what happened.

A warm September evening long foreseen,
Meeting of friends, anticipation keen,
Alumni gathering from far and wide,
Recalling youthful feats with flush of pride.
Past Senate House and King's we made our way
Intent upon the imminent soirée;
Discoursing on the feast that lay ahead
Befitting ancient Hall, High Table spread:
Potage Saint Germain from a steaming bowl,
Roast Lamb perhaps or Grouse en Casserole,
Followed perchance by Armagnac and Prune
To savour from a Georgian silver spoon;
Cheese Fritters á la Turque, and Petits Fours,
And other tasty morsels, to be sure.
As for the wines, would it be finest Rhône,
Zeltinger Himmelreich or Vintage Beaune?
Thus musing, soon to Corpus we advented,
And at the Porters' Lodge ourselves presented.
The Porter greeted us with practised eye,
Responding to our quest with gentle sigh,
Alumni Gathering? T'was last night, he said,
But would you like a cup of tea instead?

And added kindly, Have you come from far?
From Edinburgh yes, said Kay, by car,
As if this quite explained our late arrival;
In tardiness t'was plain we had no rival.

Said I, I'm sure we have some decent wine,
And I can rustle up some beans on toast;
Said Kay, that's what I'd really like the most;
So home we went and privily did dine.

Our better halves, who'd supped elsewhere that night,
Shed tears of laughter at our dreadful plight.
It always was a Wednesday, said Kay,
How could I've known the day was yesterday?
But what the hell, we'd bread and wine a-plenty,
Communion fit for Cambridge cognoscenti.

Benedicto benedicatur

Genesis: Cosmological Echoes

Dark Energy continues to be one of the great scientific mysteries of our time. This poem seeks to place it centre-stage. [Explanatory notes can be found on my personal website.]

There was a time when Time itself stood still
And Triune Space was formless, void and vast;
No matter stirred, there was no eye to see
Nor mind to comprehend the vacant past.

And yet within this carapace of calm,
In subjugation to the laws of chance,
Dark Energy lurked stealthy in the shade,
Provoking random waves in ghostly dance.

As when Earth's winds and ocean waves conspire
To focus energy in gathering storm,
In hurricane of chiral power immense,
Or maelstrom far exceeding any norm,

So these primordial space-time waves converged
With flux of energy t'ward caustic point,
Focus of pressure infinite, intense,
Where graviton and photon were conjoint.

Such fusion of extremes could scarce endure:
Explosive stress induced chaotic schism,
Releasing pent-up energy as mass
In sonoluminescent cataclysm..

There was a time when molecules converged
In replicative mode precursing Life,
Genetic coding, helices that merged,
A spiral staircase to our world of strife.

Dark Energy still roams athwart the bound
Where lightning flits and quarks have ceased to churn,
That sphere in spectral darkness all begowned,
That bourne from which no echo can return.

In this spirit, I was moved to compose an adaptation of the Apostles' Creed, that seems to me to be appropriate to the times we live in:

A Natural Philosopher's Creed

I believe in the Dark Energy that powers the Universe and all that it contains;
And in Jesus of Nazareth, mentor and inspiration of the human race, Who was conceived and born in humility;
Who suffered under the yoke of oppression, was crucified, dead and buried; he departed this life.
But hope revived within the human heart,
Infused by that hidden Energy, source of all virtue, justice and benevolence,
Whose denial leads but to desolation and decay.
I believe in the universal spirit of humanity; in verifiable scientific truth; in the Fellowship of those who seek truth; in man's enduring capacity for self-renewal; in rebirth from the dust and ashes of the stars; and in life everlasting from generation unto generation.
Amen.

G K B

On 9th May 2004, we held a Reception at DAMTP to commemorate George Batchelor, (1920--2000). I composed this sonnet to fit the occasion:

Portrait of George Batchelor, first Head of DAMTP, by Rupert Shephard

"I don't quite understand …", he used to say
In questing tone to which we lent our ears,
At Friday seminars in the old Room A,
The ones he'd never missed in fifty years;
"I don't quite understand your line of thought
That leads you to these curves of rising slope
Which fail to go to zero as they ought;
Perhaps you've lost a sign, or so I hope!"

The speaker, blanched and halted in his tracks,
Would stammer "Well, I hadn't noticed that"
(Thinking, O God, my theory's full of cracks)
"Let's leave these curves for later private chat".
So then would G K B with patient tact
Convey the insight that the treatment lacked.

Threescore Years and Ten

A two-day meeting entitled "Turbulence, Twist and Treacle" was held in April 2005 to mark the occasion of my 70th birthday, at which were gathered a wonderful array of former students and colleagues. I offered the following sonnet in my reply to the toast proposed by my good friend and colleague Nigel Weiss following dinner in the Hall of Trinity College.

> Attaining thus my threescore years and ten
> Remote from mother-land of hills and streams,
> Long settled on this damp and windswept fen
> And buried under stacks of unread reams,
> I ruminate on work of decades past,
> The depths of turbulent flows to understand,
> To pluck some pebble from that ocean vast
> Of truth that beckoned Newton from the strand.
>
> From distant realms we're gathered here to share
> Perceptions of that mobile fluid state,
> Whose universal laws hold us ensnared
> In vortex knots we still can't integrate!
> Yet, struggling so, this truth we'll soon discern:
> Through mutual teaching, we most haply learn.

Part II Limericks

In 2001, I gave (for my last time) the Cambridge 24-lecture Part II course on Fluid Dynamics. I decided to terminate some at least of these lectures with a limerick summarising their content. Here are these limericks, from which you can perhaps deduce the composition of the course!

1.
You know that in moments of stress
You tend to get tenser, not less;
But since stress is a tensor
You needn't feel denser;
It's tricky, I have to confess.

2
Stress is related to strain;
Surely that's perfectly plain!
But please don't neglect The viscous effect,
For this makes the energy drain.

3.
The equation of Navier-Stokes
Describes flow of egg-whites and yolks;
But now take the curl,
This will please any girl,
And may satisfy even the blokes.

4.
Profiles of speed parabolic
Are common in channels hydraulic,
In pumping of blood,
In sliding of mud,
And in suction of drinks alcoholic.

5.
The spreading of honey on toast
Is a task that gives pleasure to most;
But the pressure's unbounded
In corners not rounded,
So take care if you are the host!

6.
You need to be fairly astute
To track flow in a corner acute;
Though creeping and steady
There's many an eddy,
Too many by far to compute!

7
A sphere near a vertical wall
In Newtonian fluid will fall
Under modified weight
With proportional rate,
But sideways it won't move at all.

8.
When two fluids flow in collusion,
You'll observe a degree of intrusion;
Fingers appear,
The reason is clear,
But branching may lead to confusion.

9.
Production and spread of vorticity
Is a matter of utmost simplicity;
Near a boundary, say,
It can't get away,
In a flow of well-crafted complicity.

10.
Flow past a cylinder bluff --
Surely that's easy enough?
But flow retardation
Provokes separation,
And that's where the going gets tough!

11.
I'll show you as well as I'm able
That a vortex sheet is unstable;
The behaviour's erratic,
The growth rate's dramatic,
The problem gets most disagreeable!

12.
Let's study the rise of a bubble;
Some really go at the double;
With spherical cap
Some get in a flap;
You can see them without any trouble.

13.
The end of this course is now nigh,
You'll anticipate this with a sigh!
But please recollect
Whether prone or erect,
There are plenty more problems to try!

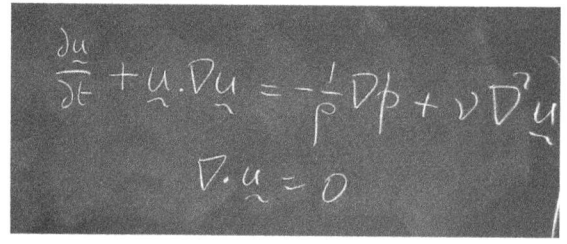

Le Château de Tennessus

Jour du Patrimoine
17th September 2005
* apologies to Chaucer (again!)

Whan that September with her mists
hath sote *
The green-brown vales and meads of vieux Poitou,
Embraced by limpid waters of its moat,
There stands four-square the keep of Tennessus;
Flanked on both sides by machicolated towers,
Whence armed and mounted knights once sallied forth,
Inspired by maidens watching from their bowers
To jousting feats that proved their valiant worth.

These ancient walls with roses now abound;
By day the massive portals stand ajar,
And by the bridge sits Tau, the faithful hound,
To welcome *voyageurs* from near and far.
As warms the château wall in morning sun,
Its gold-flecked flag proclaims proud *Patrimoine*.

Black Swan: On First Looking into the Tswaing Meteorite Crater

In single file we climbed the narrow trail
Through bramble thicket to the crater's rim,
Where warming sun cast shadows on the brim
And bathed the bushveld scrub and scattered shale.
Concealed beneath the gently waving grass
By saline lake, the haunt of duck and plover,
Lay snakes and lizards in the rain-soaked clover,
O'er diaplectic quartz and feldspar glass.

In prehistoric stone-age time of yore,
At hypersonic speed a chondrite fell
On shocked impala, thunderstruck gazelle,
And vaporised upon the forest floor.
Black swans are not so rare, I heard you say!
Beware! One may befall this very day!

Black swan on the River Cam

To Sir Nicholas Barrington on his birthday, 23rd July 2022

Thou who didst penetrate the hills of Nuristan,
Exploring far the slopes of Hindu Kush;
Who served our Government in Pakistan,
In time of Margaret Thatcher and George Bush;
Who earlier at the Chancery in Tokyo,
Held high Conciliar esteemed portfolio,
Whose diplomatic skills were well employed,
In Charge of high Affairs in old Hanoi,
At last to Cambridge town thou didst repair,
Hon. Fellow of that ancient College Clare,
To pen on vellum scroll thy life's memoir,
Distinguished climax, fitting exemplar.
Let Hafez poems infused with Shiraz wine
Inspire thee yet with soaring thoughts sublime.
On this your birthday morn, we wish you well
From friends in Banhams Close wherein you dwell.
We wish you joy along the Riverside
For years to come, with ever-bolder stride.

Sir Nicholas at Coronation Party 8 May 2023

To Bathsheba on her 14th Birthday, to accompany an illustrated copy of the Canterbury Tales, by Geoffrey Chaucer

Whan that October with his winds hath blown,
And farmers in the fields their crops have mown;
Whan that wee birdies fly to warmer climes,
And school demands that you be up betimes
And not be late and not forget your books
And coats and bags and other things on hooks
Or on the floor or somewhere p'raps upstairs
Or on the shelf or maybe under chairs;
Then when you're home from school and have some time
For tea and cakes and reading prose or rhyme,
What better words to rhyme with cup and saucer
Than all these tales composed by Geoffrey Chaucer
Six hundred years and somewhat more ago,
When pilgrimage was hazardous and slow;
When horseback was the only way to travel,
On bridleways of grass and muddy gravel?

Here anyway's a birthday gift for you,
Dear Bathsheba, with love from your . . . guess who!

Public domain portrait of Chaucer, (c.1343--1400) by Thomas Hoccleve, 1412

Address at our Golden Wedding Celebration, in response to the toast of Sandy Robertson, in the Trinity College Dining Hall, 18th December 2010, a day of biting cold and drifting snow, the coldest day so far in Cambridge in this millennium.

Great Court of Trinity College, the morning of our Golden Wedding

The Hall of Trinity College, 18 December 2010 photo: Jill Paton-Walsh

Friends, neighbours, family, Fellows and peers,
And Sandy, bosom friend from childhood years:
I warmly greet you in this lofty Hall
Where learned men look down from panelled walls,
Men of distinction, men of diverse talents,
--- And Mary Tudor here for gender balance.

I thank you all who braved this Arctic weather,
That we might gladly eat and drink together.
Of wine, I trust you have imbibèd well,
For now I have a touching tale to tell.
Forgive me if I weep to tell this tale;
'Tis but the frailty of an ageing male.

Linty was sweet thirteen when first we met
With sparkling eyes and hair as black as jet.
Springtime it was, as ever love was blind:
I vowed right then our lives should be entwined.
Thenceforth upon my soul her radiance shone
For ten long years I wooed her --- off and on!
And so it came to pass that we were wed;
"Wilt thou do this and that?" --- We will, we said.
I wore the kilt, she's worn the trousers since,
And fed me well on porridge, kale and mince.

Some months ere then, a warm September day,
I mind it well, as 'twere just yesterday!
We'd picnicked on the slopes above Loch Voil,
No better place to rest from mental toil;
A ring I had concealed within my sporran,
(This self-same pouch through many years I've worr'n)
A modest jewel from family vaults purloined,
To pledge my troth that we might be conjoined.
So having drained a flask of claret wine,
On bended knee, I pled "Wilt thou be mine?"
To which she answered "Yes, let's wilt together";
We wilted there and then upon the heather!

Soon to these Southern realms we did repair,
Escaping fine Scotch mists, --- we'd had our share;
Now blessed with children; three of them are here
   --- While Fergus tends us in another sphere.
Grandchildren furthermore attend this luncheon,
Pandora's Box the website where they function.

Well, that's the story of our life together,
O'er hills and vales, through fair and stormy weather.
The clock ticks on, and we must fain grow older,
Let's hope the climate won't get any colder!
And now, let's call upon these minstrels four,
To entertain us from this hallowed floor.

The minstrels four,
from left, Bathsheba,
Tabby, Chloe and Alfie,
December 2010

. . . and some years later,:
October 2016

And to come full circle:

A Drinker's Guide to ICTAMs past

In **Stresa** 'twas Asti Spumante,
Of that there was certainly plenty;
And then a dash of vermouth,
Of that there was ample forsooth.. 1960

In **Munich** they'd fill a huge stein
With ale of a standard divine!
1964

In **Stanford**, a pint of Budweiser,
Served well as a fine appetiser.
1968

In **Moscow** no doubt it was vodka,
Stolichnaya shipped from Kamchatka?
1972

In **Delft**, advocaat was a must,
To give drooping minds renewed thrust.
1976

In **Toronto** a cocktail of Caesar,
Just right as a thirst-quenching pleaser.
1980

In **Lyngby** we sipped Aquavit
After seminars, Oh what a treat!
1984

Grenoble, red wine of the Rhone,
A fine bottleful, vintage Beaune.
1988

In **Haifa** fresh Limonana,
To be drunk over shouts of hosanna!
1992

In **Kyoto** a cup of warm Sake,
With sushi that's really quite wacky.
1996

In **Chicago**, coke-cola with gin,
Enough to make your head spin.
2000

In **Warsaw,** it's vodka again,
Yes please, and again and again.
2004

In **Adelaide** litres of Fosters,
A brewer that beats all imposters.
2008

In **Beijing**, baijiu and tea
Was very well suited for me.
2012

In **Montreal,** caribou's tempting
All other tipples pre-empting.
2016

For **Milan**, you could drink what you like,
Zoomed at home or out on a bike.
2021

In **Daegu**, soju was the choice
That made us both laugh and rejoice.
2024

Vienna 2028: Here I come!

APPENDIX

Some Key Developments in Fluid Dynamics since 1956

JFM volumes 1-1000

It may well be argued that the 'modern era' of fluid mechanics began with the launch of the *Journal of Fluid Mechanics* (*JFM*) in 1956 by its founder-editor George Batchelor. This has been one of the most successful journals published by Cambridge University Press (CUP) during the past half-century. It has expanded continuously, and has recently celebrated the publication of Volume 1000, in December 2024. This seems a good moment to take stock of the way at least some areas of fluid mechanics (or fluid dynamics, as it is better described) have developed over the last seven decades. I can only comment on those areas in which I have been personally involved through my own research, a very small fraction of the whole immense field of study.

Magnetohydrodynamics

It was in 1957 that research in thermonuclear fusion was declassified at international level. This led to a tremendous burst of research in plasma physics and its fluid-dynamical cousin, magnetohydrodynamics (MHD). Research in MHD boomed for the next 5 years, while all the 'hanging fruit' of the subject was gathered, before settling down to an equilibrium state of progress. Cowling's 1957 monograph *Magnetohydrodynamics* gave great impetus to this field of research. He devoted a chapter to dynamo theory, i.e. to the possible generation of a magnetic field from an initial seed field by fluid motion, but this possibility was still then a matter of conjecture. The main focus of MHD research was on thermocuclear fusion devices like the tokamak, and on the interaction of the flow of the plasma in such devices with the magnetic field produced by currents in external coils. There were also pressing questions concerning the liquid metal (e.g. lithium) blanket that could in principle be used to extract heat from the plasma; the interaction of this blanket with the applied magnetic field again required detailed consideration. Through such issues, the field of MHD rapidly opened up. Such work was also relevant in solar physics and more broadly in astrophysics. It was on MHD turbulence that my own PhD research from 1958-1962 was focused.

Statistics of turbulence

The 1961 Marseille Colloquium on the *Statistics of Turbulence* was of great importance for the development of this subject. The most novel approach presented at this meeting was Kraichnan's direct-interaction-approximation (DIA) for the closure problem of turbulence which was at that time seen as quite central. The theory immediately came under fierce criticism, which Kraichnan rebutted; but this led him later in the 60s to develop his 'Lagrangian history direct interaction approximation (LHDIA)', which resolved some of the earlier issues, but was of such complexity that it has failed to attract universal support, even to this day. Batchelor was concerned at the direction research in turbulence was taking, and it is noteworthy that he himself gave up his research in turbulence from 1961 on (just as G.I.Taylor had done in the post-war years) and devoted himself instead to his influential textbook *An Introduction to Fluid Dynamics* published in 1966. He did return briefly to turbulence with his 1969 paper (Batchelor 1969) on two-dimensional turbulence, following on Kraichnan (1967), a topic that seemed artificial at the time but turned out later to have relevance for atmospheric dynamics, for which large-scale motions are constrained to be nearly two-dimensional by the Coriolis force associated with the Earth's rotation.

Geophysical Fluid Dynamics

In fact *Geophysical Fluid Dynamics (GFD)*, encompassing particularly ocean and atmosphere, was an area that developed rapidly through the 1960s and beyond. Owen Phillips was appointed an Assistant Director of Research (ADR) in GFD in DAMTP in 1961, and was succeeded in this role in 1964 by Francis Bretherton, who later went on to become Director of the National Center for Atmospheric Research (NCAR) in Boulder, Colorado. The subject expanded rapidly under their influence. Few would deny that weather forecasting now is infinitely better than it was in 1960, and this is largely the result of the global research activity in GFD carried out since the 1960s, coupled with the huge increase in computing power over that period. The foundation of the journal *Geophysical and Astrophysical Fluid Dynamics* by Paul Roberts, currently edited by Andrew Soward, has provided great support for this broad area of research.

Phillips was already well known for his pioneering work from 1957 on wind-wave interaction, as was John Miles at the Scripps Institute of Oceanography (Phillips 1957, Miles 1957). But GFD expanded in the 60s and 70s with increasing recognition of climate warming, and the effect of pollutants in the atmosphere. This stimulated research on atmospheric dynamics, wind flow over topography, and also the general ocean circulation. Problems of weather-forecasting formed a very small part of this overall area of investigation. The mechanism of hurricane formation, and the prediction of their landfall remains challenging to this day.

Dynamo Theory in Geophysics and Astrophysics

In the late 1960s and 1970s, the essential mechanism of dynamo instability due to turbulence in conducting fluids was finally understood. I am prejudiced, but I have always considered this to be one of the most dramatic advances in turbulence research since 1960. Certainly, it is now universally accepted that turbulence in a sufficiently large volume of rotating conducting fluid will always give rise to a spontaneous magnetic field, provided the turbulence is chiral, i.e. lacks reflection symmetry; and this is always the case when there is a local flux of energy parallel to the rotation vector. This was of crucial importance for geomagnetism, for which the conducting fluid is the molten iron/nickel mixture in the outer core of the Earth; and equally for astrophysics, for which the conducting fluid is the ionised gas in the convective zones of stars like the Sun. Prior to 1960, dynamo theory had been dominated by 'anti-dynamo theorems' e.g. the famous anti-dynamo theorem of Cowling (1933); post 1970, dynamo theory, taking chiral turbulence into account, had been totally transformed.

Microhydrodynamics and Suspension Mechanics

It was perhaps no coincidence that, when George Batchelor turned his attention to suspension mechanics and, more broadly what came to be called 'microhydrodynamics', this 'low-Reynolds-number' area of fluid dynamics rapidly 'took off'. Batchelor's student John Hinch, together with post-docs Gary Leal and Howard Stone were attracted into this field, which had lain dormant since Einstein's early paper (Einstein 1906); he had obtained the first-order increase in viscosity of a fluid due to the presence of a very small concentration c of suspended particles. The great difficulty was to extend this theory to higher order in the concentration c, a problem that required what was in effect a renormalisation procedure. Suspension mechanics became a research area of central importance in Chemical Engineering during the 70s and 80s.

Biological Fluid Dynamics

In parallel with this, applications of fluid mechanics to biology and medicine became equally prominent during this period. Colin Caro had founded the Physiological Flow Studies Unit (PFSU) at Imperial College London in 1966, where much work on the fluid mechanics of the cardiovascular system and the bronchial airways had been initiated. This continued with the arrival of James Lighthill in Cambridge as Lucasian Professor in 1969, and his research student John Blake; and was further strengthened when Tim Pedley returned to Cambridge from PFSU in 1973 to take up a Lectureship in DAMTP. Lighthill interacted with the zoologist Torkel Weis-Fogh in analysing the flight of birds and insects, and he carried out a major investigation on the swimming of fish and other marine organisms, stimulating research in the general field of aquatic animal propulsion. His 1975 book *Mathematical Biofluiddynamics* provides an authoritative survey of the whole field of external and internal fluid dynamics in biological systems.

Fractals and Chaos

The work of Benoit Mandelbrot on chaos theory had by 1975 achieved renown through his brilliant exploitation of computer graphics to illustrate the concept of 'fractals' that he popularised. His approach found application in fluid mechanics with his paper (Mandelbrot 1975) *On the geometry of homogeneous turbulence, with stress on the fractal dimension of the iso-surfaces of scalars*. The associated concept of 'multifractals' was later developed by Meneveau & Sreenivasan (1991) and Frisch (1991). These approaches relate to the intermittency of turbulent energy dissipation, and seek in part to provide a correct derivation of the exponents of the structure functions introduced by Kolmogorov (1962); so this resided still within the framework of statistical approaches to 'the problem of turbulence'.

Lagrangian chaos

It came more of a surprise that a simple time-periodic two-dimensional flow could have chaotic particle paths, as demonstrated in the 'blinking vortex model' of Aref (1984); such chaos is generic in such flows, and also as later described (Bajer & Moffatt 1990) in steady fully three-dimensional flows, a kinematic property present even in the limit of low Reynolds number. Lagrangian chaos is conducive to the efficient mixing of any passive scalar field, as emphasised in the book by Ottino (1990).

Direct Numerical Simulation (DNS)

Direct numerical simulation (DNS), i.e. computer simulation of Navier-Stokes turbulence without approximation (except the approximation inevitably introduced by any numerical discretisation procedure) can be traced back to Orszag & Patterson (1972). The computational expense involved increases rapidly with increasing turbulent Reynolds number (see, for example, Moin & Mahesh 1998). Nevertheless, DNS has been valuable in revealing the complex structure of vorticity at the smallest scales of turbulence, as in Ishihara et al. (2009).

Coherent Structures

Coherent structures have played an important part in the analysis of turbulent shear flows, such as boundary layers, mixing layers, jets and wakes; these are persistent vortical structures that are observed repeatedly, mainly on the larger turbulent scales of such flows. They were already observed by Leonardo da Vinci and documented in his sketches more than 500 years ago. In modern times, they are what Townsend described as the 'big eddies' in his 1956 monograph *The Structure of Turbulent Shear Flows*. The term 'coherent structures' appears in Cantwell et al (1978), who applied it to the 'turbulent spots' that appear in transition to turbulence in a boundary-layer flow; the term has been more generally used since then, even in the idealised case of homogeneous isotropic turbulence (see, for example,

Melander & Hussain 1993). Work centred on coherent vortices is not statistical in character; it involves analysis of particular interactive vortex structures evolving according to the Navier-Stokes equations.

Concentrated vortices

Concentrated vortices have the property that pressure is minimal at the vortex core, so that in a turbulent flow of water, cavitation bubbles may appear near such vortex cores. This phenomenon was exploited by Douady et al(1991), who observed the intermittent appearance of concentrated vortices in a fully developed turbulent flow. Linked and knotted vortices were created and similarly observed by Kleckner et al.(2013). These observations, and parallel results emerging from DNS, have stimulated much current work on the nonlinear interaction of vortices, as governed by the Navier-Stokes (NS) equations. This has a bearing on the 'finite-time singularity problem' (one of the million-dollar millennium problems posed by the Clay Mathematics Institute: in simple terms, do solutions of the NS equations for an incompressible fluid remain smooth for all times if smooth and of finite energy initially? Or alternatively, is there any smooth initial condition for which a singularity of vorticity will appear within a finite time? Huge effort has been devoted to this problem over recent decades, particularly by the functional analysis community, but the million dollars remains as yet unclaimed.

Nonlinear stability

Linearised hydrodynamic stability theory was already well-developed by 1960, as described in C.C.Lin's 1955 monograph *The Theory of*

Hydrodynamic Stability. But linear theory could only describe small perturbations to laminar flows, which grow exponentially if the flow is unstable. In this case nonlinear effects soon take over, either leading to saturation of the growth, or to transition to turbulence. Theoretical work on nonlinear stability theory dates from Stuart (1958), and advanced rapidly through the 1960s. It was later recognised that flows that are linearly stable could be subject to subcritical nonlinear instability. This could arise through a transient instability at high Reynolds number growing linearly for a long time till subject to a secondary exponentially-growing instability. This mechanism underlies the transition to turbulence that occurs in flows like pressure driven flow in a pipe, which are stable at all Reynolds numbers on linear theory. These aspects of stability theory have attracted increasing attention over the last 30 years.

Convective turbulence

When a fluid between two horizontal plates is heated from below it becomes unstable leading to a state of convective turbulence if the heating is strong enough. This is a subject that has been extensively studied since 1960, as recently reviewed by Lohse & Shishkina (2024). Under extreme circumstances, blobs of hot fluid erupt from the thermal boundary layer on the lower heated plate driving a state of thermal turbulence in the interior. The same sort of mechanism occurs in compositional convection driven by the release of lighter elements at the lower boundary, as in the 'mushy zone' at the boundary between the solid inner core of the Earth and the liquid outer core, where slow solidification is taking place. Compositional convection subject to global rotation ('magnetostrophic turbulence') is a favoured mechanism for the dynamo maintenance of the Earth's magnetic field, and is much studied in this context.

Free Surface Flows

Free-surface flows have attracted increased attention over recent decades, largely because of their practical importance (e.g. for inkjet printing) at the micro-scale (or even nano-scale) level, at which the relevant Reynolds number is very small. Surface tension must then be taken into consideration, and the dynamics of such flows often reflects a competition between viscous and surface tension effects, inertia being of secondary importance. Classical problems in this area include the problem of a rising bubble in a viscous fluid, as reviewed by Wegener & Parlange (1973), and the problem of the capillary instability of a liquid jet (Rayleigh 1878).

I became personally involved in this area through the observation that cusps can form on the free surface of a viscous fluid, when a converging flow is induced on the surface (Jeong & Moffatt, 1992). Normally, one would expect surface tension to smooth out such cusp-like singularities, but here it is quite feeble in competing with the effect of viscosity. The cusping phenomenon is what allows the absorption of air into a viscous fluid through the thin sheet of air created below the cusp (Eggers & Fontelos, 2015). This sort of process is crucial for the mixing of two fluids. A similar process is presumably responsible for the bubbles that form when water flows from a tap into a bath, or even when a wave breaks on the seashore; the associated high-Reynolds number problem still defies analysis.

Another familiar 'kitchen-sink' problem concerns the flow of a viscous thread of fluid from a nozzle held above a horizontal surface. The natural tendency for coiling can be overcome by moving the nozzle horizontally at sufficient speed. Many variants of this behaviour have been investigated under controlled experimental conditions (Ribe et al. 2012). Finally, problems of jet break-up and of impact of drops on solid substrates are of central importance for the process of inkjet printing (as recently reviewed by Lohse 2022), without which this book, like all others in this day and age, could not have been published in printed form.

REFERENCES

Aref, H. 1984, Stirring by chaotic advection. *J.Fluid Mech.* **143**, 1-21.

Bajer, K. & Moffatt, H.K. 1990, On a class of steady confined Stokes flows with chaotic streamlines. *J.Fluid Mech.* **212**, 337--363.

Batchelor, G.K. 1966, *An Introduction to Fluid Dynamics.* CUP.

Batchelor, G.K. 1969, Computation of the energy spectrum in homogeneous two-dimensional turbulence. *Phys. Fluids,* **12**, 233-239.

Cantwell, B. et al. 1978, Structure and entrainment in the plane of symmetry of a turbulent spot. *J.Fluid Mech.* **87**, 641-672.

Cowling, T.G. 1933, The magnetic field of sunspots. *Mon. Not. Roy. Astr. Soc.* **94**, 39-48.

Cowling, T.G. 1957, *Magnetohydrodynamics.* Interscience.

Douady, S. et al. 1991, Direct observation of the intermittency of intense vorticity filaments in turbulence. *Phys, Rev. Lett.* **67**, 983.

Eggers, J. & Fontelos, M.A. 2015, *Singularities: formation, structure, and propagation.* CUP.

Einstein, A. 1906, A new determination of molecular dimensions. *Ann. Phys.* **19** 289-306.

Frisch, U. 1991, From global scaling, à la Kolmogorov, to local multifractal scaling in fully developed turbulence. *Proc. Roy. Soc. A*, **434**, 89-99.

Ishihara, T. et al. 2009, Study of high–Reynolds-number isotropic turbulence by direct numerical simulation. *Ann. Rev. Fluid Mech.* **41**, 165-180.

Jeong, J. T. & Moffatt, H. K. 1992, Free-surface cusps associated with flow at low Reynolds number. *J.Fluid Mech.* **241**, 1-22.

Kleckner, D. & Irvine W. 2013, Creation and dynamics of knotted vortices. *Nature Phys.* **9**, 253-258.

Kolmogorov, A.N. 1962, A refinement of previous hypotheses concerning the local structure of turbulence in a viscous incompressible fluid at high Reynolds number. *J.Fluid Mech.* **13**, 82-85.

Kraichnan, R.H. 1959, The structure of isotropic turbulence at very high Reynolds numbers. *J.Fluid Mech.* **5**, 497-543.

Kraichnan, R.H. 1966, Isotropic turbulence and inertial-range structure. *Phys. Fluids*, **9**, 1728-1752.

Kraichnan, R.H. 1967, Inertial ranges in two-dimensional turbulence. *Phys. Fluids*, **10**, 1417-1423.

Lighthill, Sir James, 1975, *Mathematical Biofluiddynamics*. SIAM.

Lin, C.C. 1955 *The Theory of Hydrodynamic Stability*. CUP.

Lohse, D. 2022, Fundamental fluid dynamics challenges in inkjet printing. *Ann. Rev. Fluid Mech.* **54**, 349-382.

Lohse, D. & Shishkina, O. 2024, Ultimate Rayleigh-Bénard turbulence. *Rev. Mod. Phys.* **96**, 035001.

Mandelbrot, B. 1975, On the geometry of homogeneous turbulence, with stress on the fractal dimension of the iso-surfaces of scalars. *J.Fluid Mech.* **72**, 401-416.

Melander, M. V. & Hussain, F. 1993, Coupling between a coherent structure and fine-scale turbulence. *Phys. Rev. E*, **48**, 2669-2689.

Meneveau, C. & Sreenivasan, K.R. 1991, The multifractal nature of turbulent energy dissipation. *J.Fluid Mech.* **224**, 429-484.

Miles, J. W. 1957, On the generation of surface waves by shear flows. *J.Fluid Mech.* **3**, 185-204.

Moin, P. & Mahesh, K. 1998, Direct numerical simulation: a tool in turbulence research. *Ann. Rev. Fluid Mech.* **30**, 539-578.

Orszag, S. A. & Patterson Jr. G.S. 1972, Numerical simulation of three-dimensional homogeneous isotropic turbulence. *Phys, Rev. Lett.* **28**, 76--79.

Ottino, J.M. 1989, *The kinematics of mixing: stretching, chaos, and transport*, CUP.

Phillips, O. M. 1957, On the generation of waves by turbulent wind. *J.Fluid Mech.* **2**, 417--445.

Rayleigh, Lord, 1878, The influence of electricity on colliding water drops. *Proc. Roy. Soc.* **28**, 405-409.

Ribe, N. M. et al. 2012, Liquid rope coiling. *Ann. Rev. Fluid Mech.* **44**, 249--266.

Stuart, J. T. 1958, On the non-linear mechanics of hydrodynamic stability. *J.Fluid Mech.* **4**, 1-21.

Townsend, A.A. 1956, *The Structure of Turbulent Shear Flow*. CUP.

Wegener, P. P. & Parlange, J.Y. 1973, Spherical-cap bubbles. *Ann. Rev. Fluid Mech.* **5**, 79-100.

INDEX

This index is arranged in the following sections:
FAMILY
PEOPLE
PLACES
POETRY
RESEARCH TOPICS
SCIENTIFIC ORGANISATIONS

FAMILY

Chloe, Tabitha, Alfie, Bathsheba
    grandchildren, 79, 134, 135, 157, 204, 207
Emmeline, mother, 24, 178
    death in car crash, 117
    in China, 91, 92
    in Strawberry Hill, Jamaica, 51
Fergus, son, 9, 38, 45, 47, 50, 51, 205
    death, 97
    manic episode, 86, 87
    miracles, 99, 171
Fred, father, 24
Fred Tingey, son-in-law, 79, 98, 134
Granny Fleming, 25, 51, 139
Hester, daughter, 9, 50, 66, 98, 135, 154
    in Maisons Laffitte, 134, 135
    in Thessalonika, 97
    lymphoma crisis, 79
Iain, Jamie, Kennie, David, Eleanor, Peter
    Canadian cousins, 19, 48
Kay Stiven, brother-in-law. and Sue
    Dinner in Corpus, 190--191
Lindesay, sister, 24, 25, 33, 126, 177
Linty, wife, vii, 15, 16, 27, 45, 46, 47, 48, 67, 68, 79, 88, 110, 114, 115, 116, 122, 123, 135, 144, 147, 148, 149, 154, 178, 189
    marriage, 38, 205--207
    twin brother, Bobby, 116
    and Marlene, 48
Penelope, daughter, 50, 67, 85, 88, 95, 154
    in Daegu, 20, 21
    in Susa, 87
Peter, son, 40, 45, 47, 50, 51, 161
    Giffen goods, 161, 162
Phoebe, Aunt (and Uncle John), 19, 48
Stiven, Rev. Dr. David, father-in-law, 70

PEOPLE

Abarbanel, Henry, 95
Acrivos, Andy and Jenny, 107, 157
Adrian, Lord, 50, 54
    and Hester, Lady Adrian, 50
Aitken, A.C., 27, 166
al-Marashi, Ibrahim, 144
    dodgy dossier, 144

Anderson, Eric
  Headmaster of Eton, 26
Aref, Hassan, 10, 11, 13, 15, 17, 18, 95,
  122, 142, 181, 216, 223
Arnol'd, V. I. (Dima), 64, 82, 84, 88, 89,
  90, 100, 134
Assange, Julian, 174
Atiyah, Sir Michael, 105, 106, 109, 117,
  118, 149, 167
Lily, Lady Atiyah, 167

Babbage, Dennis, 29
Bajer, Konrad, 15, 100, 128, 131, 146,
  154, 156, 165, 216, 223
  and Margołzata, 154
  death in 2014, 160, 161
  funeral in Warsaw, 161
Barenblatt, G.I. (Grisha), viii, 8, 80, 83,
  84, 95, 107
  in Stresa, 3
  in Moscow, 80
  in Munich, 5
Barrington, Sir Nicholas, 203
  former Ambassador to Pakistan
Batchelor, G.K.(George), viii, 2, 7,
  12, 14, 32, 33, 38, 39, 40, 41,
  44, 66, 74, 78, 94, 126, 150,
  211, 212, 213, 215, 223
  70th birthday, 106, 107
  centenary, zoomed, 173
  Batchelor Prize, 19, 145
  death in 2000, 125
  early retirement, 84
  J. Fluid Mech., 49
  sonnet to GKB, 195
  Treasurer of IUTAM, 3
  Wilma and daughters, 42, 43
Benedetti, Nicola, 157
  violinist from Scotland
Berry, Sir Michael, 123
Besicovitch, Abram, 5
Betchov, Robert, 54
Blackwell, Chris, 24

  Founder of Island Records
Blake, John, 215
Bogoliubov, N., 3
Bolcato, Robert, 95
Boley, Bruno, 8, 9
Bondi, Hermann, 129
Born, Max, 28
Bragg, Laurence, 50
Braginski, S.I. (Stanislav), 82
Brancher, Jean-Pierre and Agnès, 73
Brandenberg, Axel, 117
Branicki, Michal, 128, 130
Branover, Herman, 70
Bray, Francesca, 144
Bretherton, Francis, 143, 213
Brezhnev, Leonid, 80
  Former President of USSR
Broers, Sir Alex, 120
Brooke Benjamin, T., 42, 45, 107
Buckeridge, John, 18
Budiansky, Bernard, 3
Bullard, Sir Edward (Teddy), viii, 48,
  62
Buniy, Roman, 156
Burbanks, Andrew, 129, 130, 142
Burgers, J.M., 1
Burns, Robert, 168
  Scotland's national poet
Bush, George, 205
  Former President, USA
Bush, George W., 144
  Former President, USA
Bush, John, 136
Butler, R. A. (Rab), 54, 55
  Former Master of Trinity College

Cantarella, Jason, 104
Cantwell, B., 223
Carbonaro, Pantaleo, 88
Caro, Colin, 215
Carrier, George, 3
Cartwright, Mary, 167
Chen, Louis, 141

Childress, Steve, 174
  75th birthday, 153
Chong Chi Tat, 141
Chong, Min, 133
Chui, Atta, 13, 104
Coates, John, 130
Comte, Pierre, 85, 95, 154
Corrsin, Stan, viii, 49
Couder, Yves, 135, 136
  bouncing droplets, 135
  wave-particle duality, 136
Cowling, T.G., 212, 214, 223
Craik, Fergus, 26, 30, 31
Crighton, David, 10, 11, 107, 126, 133
  death in 2000, 125
  Head of DAMTP, 107
  Master of Jesus, 125

Da Rios, L.S., 54
Dasgupta, Partha, 146
de Gennes, Pierre-Gilles, viii, 66
de Rothschild, Sir Evelyn, 118
Dellar, Paul, 13
Deng Xiaoping, 91
den Hartog, J.P., 3
Diana, Princess, 115
Dillon, Rick, 6
Dirac, Paul, 39, 90
Dormy, Emmanuel, 165, 174, 176
Douady, S., 218, 223
Drazin, Philip, 75
Drobyshevski, E.M., 83
Duffy, Brian, 75

Edge, W.L., 28, 38
Eggers, Jens, 146, 220, 223
Einstein, Albert, 138, 149, 186, 215, 223
Elder, John, 48
Elizabeth, HM The Queen, 120
El Sawi, Mohamed, 69
Eltayeb, Ibrahim, 69, 145, 169
Engelbrecht, Juri, 11, 15, 16

Euler, Leonhard, 37, 52, 83, 88, 126, 127, 155
Evans, David, 75

Falcone, Giovanni, 88
Farge, Marie, 148, 154, 165
Faulkes, Dill, 119, 150
Favre, A., 3, 39
Feynman, Joan, 148
Feynman, Richard, viii, 90, 91, 149, 166
  Dirac Memorial Lecture, 90
  O-rings, 91
Fiszdon, Wladek, 14, 42
Fitzgerald, Edward, 179
Fleck, Norman, 21
Fleming, Robin, 121
Fölmer, Hans, 141
Fontelos, M.A., 220, 223
Freedman, Michael, 96
Freund, I.B. (Ben), 11, 16
Friedman, Anver, 141
Frisch, Uriel, 87, 100, 109, 110, 146, 147, 216, 224
Fry, John, 130
Fry's Electronics, 130, 131

Gailitis, Agris, 83, 98
Gama, M. et Mme, et Annie, 26
Germain, Paul, 2, 9, 17
Germano, Massimo, 54
Gershon, Peter, 150
Gilbert, Andrew, 153
Glennie, Charles, 28, 29
Goldstein, Ray, 19, 164
Golitsyn, Georgi, 47, 83
Gorbachev, Mikhail, 109
  Former President, USSR
Greene, John, 108
Guest, Howard, 173, 175
Guyon, Etienne, 66
Hahn, Fritz, 140
Hamson, Jack, 38, 55

Hasimoto, H., 100
Hawking, Stephen, viii
　arrival in Cambridge, 40
　Brief History of Time, 86, 103
　Leverhulme support, 85
　Lucasian Chair, 77, 85
　and Jane, 86
　voice synthesis hardware, 86
Hayes, Michael, 11, 12
Haynes, Peter, 150
Hellstedt, Anne, 133
Helmholtz, Hermann von, 52
Heraclitus, 110
Hercynski, Andrzej, 157
Hercynski, Jannick, 157
Hercynski, Ryczard, 14, 42, 157
Hiddleston, Jim, 26, 79
Hinch, John, 67, 146, 215
H.M. King Charles III, 55
Hoff, Nicholas, 5
Hornung, Hans, 132
Howarth, Leslie, 74
Howe, Roger, 141
Hoyle, Fred, 39, 40
　exodus of, 44
　steady-state theory, 129
Huerre, Patrick, 114
Hughes, David, 169
Hunt, Julian, 146
Hunt, Robert, 122
Huppert, Herbert, 175
Hussain, Fazle, 47, 100, 168, 218, 225
Huxley, Sir Andrew, 105
　Former Master of Trinity College

Imai, Isao, 3
Irvine, William, 224
Ishihara, T., 217, 224
James, Eric, 99
　Chaplain of Trinity, 171
Jarvis, Dick, 2
Jeffreys, Bertha, 108
Jeong, JaeTak, 11, 110, 220, 224

Jimenez, Javier, 116
Jones, Chris, 148
Jones, Gareth, 71
Jones, Vaughan, 110
Joseph, D.D. (Dan), 143

Kambe, Tsutomu (Tom), 12, 16, 111
Karman, Theodore von, viii, 1, 2, 3, 39
Kayyám, Omar, 179
Keller, J.B.(Joe), 3, 129
Kelley, David, 73
　Baudelaire, 73
Kelvin, Lord, 52, 123, 150, 167
Kemmer, Nicholas, 28, 36
Kephart, Tom, 156
Kessler, John, 96
Kida, Shigeo, 12, 110
Kimura, Y. (Yoshi), 13, 19, 156, 164, 166, 167, 176
Kingman, Sir John, 119, 120, 121
Kiya, Masuro, 111
Kleckner, D., 218, 224
Kluwick, A., 16
Knops, Robin, 75
Koiter, Warner, 6, 7
Kolmogorov, A.N., viii, 39, 83, 165, 216, 224
Kovasznay, Leslie, viii, 39
Kraichnan, R.H. (Bob), 49, 212, 213, 224
Krause, Fritz, 7, 58, 59, 62, 63
　and the STASI, 63
Kruskal, Martin, viii, 114, 181, 184
Küchemann, Dietrich, 107
Kulsrud, Russell, 114

Ladyzhenskaya, Olga, 64, 83, 101
Lagerstrom, Paco, 3
Lamb, Horace, 17
Lanchon, Hélène, 73
Lathrop, Daniel, 148
Leal, Gary, 13, 215
Lee Hsien Loong, 141

Former PM, Singapore
  supervision by candlelight, 141
Leighton, Angela, 186
Leung, Ka Hin, 141
Lickorish, Raymond, 110
Lielausis, Olgerts, 83
Liepman, Hans, 3
Lighthill, M.J. (Sir James), viii, 7, 9, 10,
  13, 17, 37, 71, 85, 91, 108, 150,
  151, 215, 224
  and Nancy, 7, 8
Lin, C.C., 94, 218, 224
Linden, Noah, 122
Linden, Paul, 132
Lohse, Detlef, 219, 221, 225
Loitsianski, L.G., 3
Long, Peter, 171, 172
Loper, David, 107, 116, 145
Lui, Pau Chuan, 141
Lundgren, Tom, 143
Lungu, Edward, 140
Lynden-Bell, Donald, 34
  Bionassay, 34
  hit cow in Jura, 35
  joint paper 2015, 163
Lyttleton, Ray, 4

Macdonald, Donald, 30, 31
Mahesh, K., 217, 225
Mak, Vincent, 13
Mandelbrot, Benoit, 216, 225
Marley, Bob, 24
Marrian, Denis, 73
Maslowe, Sherwin, 95
Massingue, Venâncio, 18
Maxwell, James Clerk, 54, 103, 149,
  167, 185--188
McCalla, Clem, 56
Melander, M.V., 218, 225
Meleshko, Slava, 17, 19
Meneveau, Charles, 225
Mercurio, Salvio, 88
Miles, John, 214, 225

Miloška, 158, 176--178
Mirrlees, Jim, 29
  Nobel Prize in Economics, 29
Mizerski, Krzysztof, 19, 148, 149
Moin, Parviz, 217, 225
Moore, Derek, 69
Moore, Gordon, 86
  Intel, 86
Moreau, Jean-Jacques, 53
Moreau, René, 70, 94, 146
Morrison, John Todd and Jessie, 139
Munk, Walter and Judith, 48
Mushkelishvili, Nicoloz, 3

Naghdi, Paul, 3
Naghib, Ahmed, 143
Narasimha, Roddam, 130
Nash, John, 146
  A beautiful mind, 146
  Nash equilibrium, 146
Newell, Alan, 123
Newton, Isaac, 78, 79, 183, 189, 195
Nietzsche, Friedrich, 43
Niordson, Frithiof, 8
Nisbet, Andrew, 28
Nott, Sir John and Miloška, 158, 176,
  177
  Here today, Gone tomorrow, 158
  Times obituary, 176
Novikov, Evgenyi, 83

Obama, Barack, 157
  Former President, USA
Oberlack, Martin, 21
Obukhov, A.M., 47, 83
Okhitani, Koji, 111
Olhoff, Neils, 8, 16
Orszag, Stephen, 67, 217, 225
Oruba, Ludivine, 165
Ottino, Julio, 143, 217, 225
Owada, Hisashi, 12

Parker, E.N. (Gene), viii, 59, 100, 117

Parlange, J.Y., 220, 226
Pascal, Blaise, 133, 134, 165
Patterson, G.S., 217, 225
Pearson, J.R.A.(Anthony), 4, 138
Pedley, T.J. (Tim), 19, 133, 146, 215
Peregrine, Howell, 75, 123
Pesci, Adrianna, 19, 164
Philip, HRH Prince, 109, 110
Phillips, O.M. (Owen), 3, 49, 107, 213, 214, 225
and Mearle, 49
Phipps, Simon, 171
Polkinghorne, John, 85, 150
Poole, Adrian, 186
Porteous, Ian, 28
Proctor, Michael, 8, 146, 169
and Julia, 146
Proudman, Ian, 40
Pryor, Mark, 32
Pucknachov, Tanya, 81, 89
Pucknachov, Vladek, 68, 81

Rädler, Karl-Heinz, 7, 58, 59, 62, 63
Ramkissoon, Harold, 145
Rangwala, Glen, 144
dodgy dossier, discovery, 144
Rank, J. Arthur, 30
Rayleigh, Lord, 54, 220, 226
Redpath, Theo, 64
Rees, Martin, 3
Former Master of Trinity College
Renner, Bruno, 58
Ribe, N.M., 221, 226
Ricca, Renzo, 13, 19, 131, 146, 163, 165, 173, 180
Richard, Alison, 150
Former Vice Chancellor, Cambridge
Roberts, Glyn, 58
Roberts, Paul, 58, 59, 82, 215
Robertson, Sandy, 144, 205
Rothschild, Lord Victor, 118
Rothschild, Nathan Meyer, 118, 119
Rothschild, Sir Evelyn de, 118

Roulstone, Ian, 119
Rüdiger, Günther, 62, 63
Ruzmaikin, Sasha, 82, 101, 109, 148, 149
Ryan, Desmond and Mary, 73

Sadourny, Robert, 116
Saffman, Philip, 107, 175
Sagilde, Sarifa, 18
Salam, Abdus, 68, 146
Salençon, Jean, 15
Sano, Osamu, 13
Satchell, Stephen, 78
Satrapi, Marjane, 135
Sawyer, John, 26
Scheihlen, Werner, 11
Schlapp, Robin, 28
Schlichting, Hermann, 3
Schneider, Kai, 154
Schrefler, Bernhard, 165
Sciama, Denis, 40
Seal, Anil, 162
Sedov, Leonid, 3, 7
Shercliff, Arthur, 8, 52, 70, 71
Shimomura, Yutaka, 127, 128, 146
Shishkina, O., 219, 225
Shuckburgh, Emily, 142
Siegmund, David, 141
Singh, Kiran, 130
Sobral, Yuri, 130
Sokoloff, D.D., 82
Solonnikov, V.A., 83
Soward, Andrew, 109, 110, 146, 174
Editor, GAFD, 175, 213
Spiegel, Ed, 78, 155
and Barbara, 155
Squire, Sarah, 15
Former Ambassador to Estonia
Sreenivasan, K.R. (Sreeni), 145, 216, 225
Stasiak, Andrzej, 156
Steenbeck, Max, 7, 58, 59
Stewart, Ian, 168

Stix, Michael, 59
Stoddart, Sandy, 149
  Sculptor, Maxwell statue
Stokes, Sir George Gabriel, 123, 150
Stone, Howard, 145, 215
  Batchelor Prize 2008, 145
St Pierre, Martin, 116
Stuart, J.T. (Trevor), 3, 219, 226
Sumners, De Witt, 173

Tabeling, Patrick, 114
Tablas, Antonio, 50, 68
  Lanestosa, Spain, 50
  Schrödinger's cat, 50
Tait, Peter Guthrie, 167
Tan, Dr Tony, 140, 141
  Former Deputy PM, Singapore
Tatsumi, Tomomasa, 11, 12, 111
Taylor, G.I. (Sir Geoffrey), 1, 3, 30, 74, 175, 213
Taylor, John, 85
Thatcher, Margaret, 77, 158, 205
Thomas, Dylan, 170
Tichborne, Chidiock, 97
Tobias, Steve, 148
Tokieda, Tadashi, 128
Toomre, Juri, 41, 146
Townsend, Alan, 217, 226
Tranah, David, 165
Truesdell, Clifford, 4, 49
Tsinober, Arkady, 100
Tuck, Ernie, 17
Turok, Neil, 136, 138
  Founder of AIMS, 15, 18
Tvergaard, Viggo, 19
Ursell, Fritz, 17

Vainshtein, Sam, 81
van Campen, Dick, 15, 16, 19
Van Dyke, Milton, 3, 107
  in Stanford, 6
Van Hove, Leon, 34
van Wijngaarden, Leen, 6, 7, 11

Vladimirov, Volodya, 13, 146
  and Natasha, 170
  in Novosibirsk, 81
  in Hong Kong, 117
  in Oman, 169

Waechter, Trevor, 4
Waleffe, Fabian, 143
Wang Ren, 9, 11, 91, 94
Wedderburn, Bill and Nina, 177
Wegener, P.P., 220, 226
Weir, Tony, 65
Weis-Fogh, Torkel, 215
Weiss, Nigel, 147, 195
Wenrui Hu, 92, 94
Whitham, Gerry, 3
Whittaker, Edmund, 167
Wiegmann, Paul, 160
Wilson, Harold, 54
Winter, Greg, 162
  Former Master of Trinity
Wittgenstein, Ludwig, 171, 172
Wolfram, Stephen, 14
Woltjer, L., 52, 53

Yaglom, Akiva, 83, 107
Yang, Philémon, 159
  Former Prime Minister of Cameroon
Yau, Jie, 168
Yih, Chia-Shun, 32
Yosinobu, Yamamoto, 111

Zakharov, Vladimir, 181
Zaslavsky, Alex, 16
Zel'dovich, Ya. B., 64, 81, 82, 89
Zheng Zhemin, 9, 16, 91, 92
Zhou PeiYuan, 94, 132
Zomahoun, Thierry, 158
Zorski, Henryk, 3
Zweibel, Ellen, 148

PLACES

Andorra, 67

Australia
  Adelaide
    ICTAM 2008, 16--18, 145, 209

Austria
  Graz, 99
  Insbruck, 99
  Vienna, 42, 99, 108, 210

Botswana
  Gabarone, 140

Brazil
  Rio de Janeiro, 15

Cameroon
  AIMS in Limbe, 159, 160
  Bobende, 160
    black sands, 160
  Yaounde, 158, 159

Canada
  Comox, Vancouver Island, 116
  Montreal, Quebec
    ICTAM 2016, 19, 209
  Ottawa, Ont., 48
  Toronto, Ont
    ICTAM 1980, 7, 208
  Whistler, BC, 116
    SEDI, 116
  Winnipeg, Manitoba, 31
    Trans-Canada pipeline, 31

China
  Beijing, 92, 94
    ICTAM 2012, 10, 18, 19
  Chengdu, 92, 93
  Hangzhou, 92
  Hong Kong, 116
  Shanghai, 92, 93
  Suzhou, 15
  Xian, 92, 94

Côte d'Ivoire
  Abidjan, 159

Czechoslovakia (former)
  Bratislava, 42
  Prague, 42

Denmark
  Bornholm Island, 32
  Lyngby ICTAM 1984, 8, 208

England
  London
    IC(T)AM 1948, 2
    Royal Society, 129
    Exhibition 2007, 129
    Humphry Davy Lecture, 124
  Bristol, 74
    Château Haut Belgrave, 74
    Upper Belgrave Road, 74
  Cambridge
    Barton Road, 46
      elderflower champagne, 46
    Chair of Mathematical Physics, 45, 78
    Chedworth Street, 46
    Churchill College, 46
      students from, 45
    DAMTP
      bubble tube, 85
      Free School Lane, 40, 41, 90
      Friday seminars, 188
      Press site, 45
    DPMMS, 110, 130
    Hammonds, 38
    Isaac Newton Institute (INI), 106, 108, 109
      boules court, 122, 181
      Director of, 117

Posters in the Underground, 121
  Queen's Anniversary Prize, 120
  Jesus College, 126
  Lucasian Professor, 39, 71, 77, 85, 209
  St John's College, 106
  Trinity College, Cambridge
    80th birthday dinner, 162, 163
    Admission of Women, 55
    Appointed Senior Tutor, 71
    Appointed Tutor, 64
    Burns Night Dinner
      Selkirk Grace, 168
    College Meeting, 56, 57
    Combination Room buzz, 163
    Gate Hours, 57
    Great Gate, 33, 57, 78, 79
    IUTAM Symposiun 1989, 100
    Yeats prize essay, 94
  Newcastle
    Soward's 80th Birthday, 174, 175
  Oxford, 88
    BTMC 1968, 37
    Exeter College, 26, 79

Estonia
  Tallin, 15
    British Embassy, 15

Finland
  Kittila, 117

France
  Cargèse, Corsica, 114
  Grenoble, 53
    ICTAM 1988, 9--10, 96, 209
    INPG, 94--95
    MADYLAM, 94
  Les Houches, Haute-Savoie, 34, 68, 69
    Aiguille du Bionnassay, 34
  Marseille
    Anniversary. Conference 2011, 154

  Turbulence Colloquium 1961, 155
  Paris
    Chaire Blaise Pascal, 133
    Denis Diderot University, 134
    Ecole Normale Supérieure, 134
    Ecole Polytechnique, 114
      ceremonial sword, 115
    LadHyX, 114
    IC(T)AM 1946, 2
    Institut Henri Poincaré, 134
    l'Hotel des Grandes Ecoles, 148
    Lycée Henri IV, 114
    Maisons Laffitte, 135
    Place Jussieu, 73
    Rue du Pot de Fer, 11, 134
    Rue Molière, 73
    Université de Tous les Savoirs, 147
      UTLS Lecture, 147
    Université Pierre et Marie Curie, 73
  Nice, 87
    Observatoire, 87
  Poitiers, 154
  St. Hilaire du Touvet, 95
    Dent de Crolles, 95
  Toulouse, 144
  Wimereux, Normandy, 144

Germany
  Herrenalb, 7
  Lubeck, 32
  Munich
    ICTAM 1964, 3, 208
  Potsdam, 7, 58, 59, 62, 63
    Declaration 1945, 63
    Sanssouci Palace, 63

Ghana
  Accra, 158
  AIMS in Biriwa, 158, 159, 160
  Cape Coast Castle, 159

Greece
  Lefkada, 98

Patras, 97
Thessalonika, 97

Hungary
Budapest, 153
Gellért thermal baths, 153

India
Bangalore, 130
Mumbai, 129
TechFest 2008, 129, 130

Iran
Isfahan, 132, 133
Persepolis, 133
Shiraz, 133

Ireland
Skreen, Co.Sligo, 123

Israel
Beersheva, 70
Dead Sea, 71
Haifa, 7
ICTAM 1992, 10, 11, 209
Jericho, 71
Jerusalem, 70
Masada, 71
Sea of Galilee, 71

Italy
Milan
Zoomed ICTAM 2021, 20, 173, 210
Mondello, 88
Palermo, 88
Universita degli Studi, 88
Perugia, 75
Segesta, 88
Stresa IC(T)AM 1960, 1, 2, 3, 208
Susa, 87
Trieste, 68
40th Anniversary, ICTP, 145

Turin, 128
Pannetti-Ferrari Prize, 128
Venice, 164, 180
Palazzo Franchetti, 164

Jamaica
Irish Town
Strawberry Hill, 24, 51

Japan
Kyoto, 110, 111
croquet in, 112, 113
ICTAM 1996, 11, 209
RIMS, 110
Kyushu, 111
Fukuoka University, 111
Tokyo, 111
Sapporo, 111
Hokkaido University, 111

Korea, S.
Daegu, 176
ICTAM 2024, 20, 21
Taekwondo, 21
Seoul, 20

Latvia
Riga, 80, 83, 98
Salaspils, 83

Malaysia
Kuala Lumpur, 142

Mexico
Tijuana, 48

Mozambique
Maputo, 18

Netherlands
Delft, 1, 5, 6, 7, 208
IC(T)AMs 1924 and 1976

Oman
  Muscat, 168, 160
  Sultan Qaboos University, 168

Poland
  Auschwitz, 43
  Białowieża, 64
  Krakow, 43
  Warsaw, 154, 161
    ECT13, 154
    ICTAM 2004, 14, 15, 17, 209
  Zakopane, 43, 131
    KGB minders, 43
    limericks 2001, 131

Russia
  Akademgorodok, 81
  Irkutsk, 80
    Lake Baikal, 81
  Moscow, 80, 88, 98
    ICTAM 1972, 6, 208
    KGB minders, 44, 82, 89
  Novosibirsk, 80, 81
  St Petersburg (Leningrad), 80

Scotland
  Aberfeldy, 30, 31
    grouse beating, 30
  Edinburgh
    Ferguson's Rock, 30
    George Watson's Boys' College, 26, 179
    Scottish Leaving Certificate, 80
    Portobello, 27
    Tait Professor, 28
    University, 126
      Whittaker Colloquium, 166
    Walls' Ice Cream, 29
    World War II, 24
      evacuated to West Linton, 24
  Inveresk, Midlothian, 27, 33
    MG TC4, 33, 34, 35
  Tomintoul, 169

  Bailebeg, 169
  Gaulrig, 117, 160
    illicit still, 169
    Glenlivet distillery, 169

Senegal
  AIMS in Mbour, 155
  Dakar, 155, 156

Singapore
  Nat. Univ. of Singapore, 140
  IMS, Singapore, 140

South Africa
  Capetown, 156
  Muizenberg, 138
  Stellenbosch, 139

Spain
  Barcelona, 174
  Madrid, 116
  Mieres, Catalonia, 144

Sudan
  Khartoum, 69
  Omdurman, 69
  Wad Madani, 69

Sweden
  Stockholm ICAM 1930, 2

Thailand
  Bangkok, 111

Trinidad and Tobago
  CCOFD, 145

Ukraine
  Kiev, 89

United Kingdom
  see under **England** and **Scotland**

USA
Ann Arbor, 143
  University of Michigan
Atlanta, Georgia, 167
  APS DFD meeting, 167
  Otto Laporte lecture, 167
Baltimore, Maryland, 39
  Johns Hopkins University, 39
Boulder, Col. 41
  NCAR, 143
Bozeman, Montana, 153
  Yellowstone National Park, 48, 153
Cambridge Mass ICAM 1938, 2
Chicago
  ICTAM 2000, 15, 95, 126, 143, 181, 209
  Midwest tour, 142
  Northwestern, 142
Illinois
  Institute of Technology, 142
  Urbana-Champaign, 142
Indiana
  Notre Dame University, 142
  Purdue University, 142
La Jolla, 47, 95
  IGPP, 95
  INLS, 95
Minnesota, 142
  University of, 142
National Academy of Sciences (NAS), 157
  elected 2008, 157
Providence, RI, 16
San Diego, 95
Santa Barbara
  KITP, 105, 108, 148
Seattle, 59
  Boeing conference, 59
Stanford, 46
  Alvarado Row, 46, 47
  Edsel Citation convertible, 46, 48
  IC(T)AM 1968, 5, 208
  Old Preservation Hall Band, 5

SS United States, 46
Tallahassee, Florida, 107
  St George Island, 108
Tucson, Arizona, 96

Wisconsin, 142
  University of, 142
Woods Hole, 16

POETRY

A Natural Philosopher's Creed, 194
Black Swan, 202
Dinner at Corpus, 190--191
Drinker's Guide to ICTAMs past, 208--210
Genesis, Cosmological Echoes, 172, 192--193
GKB, sonnet, 195
Golden Wedding Celebration, 205--207
Le Château de Tennesus, 201
Natural Philosopher's Creed, 200
Newton's Apple Tree
  planting of, 189
Part II limericks, 197--200
Tam-day musings, 181--185
The Genius o' Glenlair, 186--188
The Twa Corbies, 162, 163
Threescore Years and Ten, 147, 196
Tichborne's Lament, 97
To Bathsheba on her 14th Birthday, 204
To Sir Nicholas on his birthday, 203

RESEARCH TOPICS

corner eddies, 146
dynamo theory, 53, 58, 59, 62, 153, 176, 206, 208
Euler's disc, 155

fast dynamo, 81
free surface cusps, 11
Hele-Shaw cell, 175
   fingering instability, 175
helicity, 7, 37, 52, 53, 61, 64, 96, 104, 146, 164, 170
   mirror symmetry, lack of, 37, 61
knotted vortices, 37
MHD turbulence, 212
magnetic relaxation, 88, 96
mean field electrodynamics, 58, 59
Möbius strip soap film, 163, 164
Navier-Stokes equations, 61, 123, 165, 197, 211, 212
rattleback, 155
rising egg, 155
syrup rings, 68
tokamak, 206
topological fluid dynamics, 156
transient instability, 47, 213
turbulence, 47, 49, 53, 58, 59, 61, 67, 70, 87, 94, 95, 109, 111, 114, 116, 146, 154, 155, 156, 165, 166, 206, 207, 208, 211, 215

XCC (Exec. Comm. of CC), 7, 8, 9, 12, 13, 15, 17
ICREA (Catalan Inst.Res.Adv.Studies), 174
ICTP, Trieste, 145
ICSU (Int. Council for Science), 137
   Mozambique 2008, 18
   Rio de Janeiro 2004, 15
   Suzhou, China 2006, 15
IGPP, La Jolla, 95
IMU (Int. Mathematical Union), 137
IUGG
   GA, Vienna, 108
IUTAM, 1--20
   Bureau, 11--17, 130, 164
   elected President, 126
   General Assembly, 8--20
NAS (National Academy of Sciences}, 157
NATO Adv. Study Inst., 62
RS (Royal Society), 80, 89, 114, 122, 124, 125, 128, 129
RSE (Royal Society of Edinburgh), 166

SCIENTIFIC ORGANISATIONS

AIMS (African Inst. Math. Sci.), 136
   AIMS-NEI, 138
   capacity building, 15, 137
APS (American Physical Society), 167
   Fluid Dynamics Prize, 167
   Otto Laporte Lecture, 167
EUROMECH, 107
GEOTOP-A, 173
ICTAM, 1--20, 95, 96, 126, 145, 173, 175, 181, 208
   IPC (Int. Papers Comm.), 6, 7, 10
   LOC (Local Organising Comm.), 12, 13

www.ingramcontent.com/pod-product-compliance
Lightning Source LLC
Chambersburg PA
CBHW070141100426
42743CB00013B/2789